# Praise for SUPE

'Crisp, clever graphics, symbols and examp
He challenges the way we think and resist change, even when the alternative is disaster. *Supercrash* is a hugely readable, revelatory condemnation and call to arms.'
*Independent*

'A remarkable read, visually clever and inventive…eminently readable. He shows all sides of a problem in a way the reader can understand, while never losing track of the human aspect in the complex issues involved… Darryl's unique comic art takes the reader easily into the heart of complex matters that have important influences on everyone's lives and makes them understandable visually as well as with words.'
*Forbidden Planet*

'Cunningham's pithy prose and funky art tell a complex, important tale. He connects the dots from Ayn Rand to Alan Greenspan to the mess we're in today, tackling tangled subjects with clarity and zing.'
Michael Goodwin, author of *Economix*

'Remarkable, informed, accurate and incisive… At last there is a single, readable, beautiful book that explains to the generation that came of age after 2008 what happened, why it happened, and why it will happen again.'
Danny Dorling, author of *Inequality and the 1%*

'A provocative, thoughtful, visual essay that tackles the language of ideas. *Supercrash* will leave you better informed and, more than that, it will leave you angry.'
Teddy Jamieson, *Herald Scotland*

'*Supercrash* is nothing short of Darryl's masterpiece.'
*Bradford Telegraph & Argus*

'Decodes the ideas with virtuosity…stylish and effective text, supported by simple outlines and flat colours, plays with the rhetoric of graphic symbols.'
*L'Humanité*

'A truly outstanding piece of work, illustrated so intelligently in his wonderful no-nonsense informative style.'
*Page 45*

# Praise for SCIENCE TALES

'Deals with some of the most urgent debates in science…sorting facts from fiction and presenting complex information in a highly accessible way.'
*Observer*

'Cunningham's charming artwork complements his concise arguments…his stark lines and simple layouts give his comic the feel of a scientific analysis. The artwork is uncluttered, leaving little to distract the reader from the exposition, delivered in stripped-back, staccato prose.'
*New Scientist*

'He has managed to distil the arguments into a wonderfully clear and concise form…a great primer for those seeking arguments to undermine their *Daily Express*-reading uncle.'
*Herald Scotland*

'Makes complex issues simple, but never simplistic.'
*Headline Environment*

'Manages to be full of expression and also very evocative. Even if you haven't tried to read something in comic format before, you'll find this easy to follow on the one hand, and thought-provoking on the other.'
*Bradford Telegraph & Argus*

'Cunningham's art has clean lines and a continuity that is often graceful, charming and endearing. He speaks with quiet authority on his subjects, but is careful to cite a whole range of sources and research papers.'
*Independent*

'Brilliantly presented, and customarily classy…Cunningham delivering his message with style, great art, even moments of outright comedy.'
*Forbidden Planet*

'Both succinct and substantive, and a fierce and intelligent promoter of the scientific process over blind superstition and baseless supposition.'
*Broken Frontier*

'An eye-catching way to get across the important message that a science-based approach to understanding makes far more sense than one that is evidence-free. Cunningham draws out the fictions and lays bare the facts.'
*Chemistry World*

DARRYL CUNNINGHAM

# GRAPHIC SCIENCE

## SEVEN JOURNEYS OF DISCOVERY

myriad

First published in 2017 by

Myriad Editions
New Internationalist
The Old Music Hall
106–108 Cowley Rd
Oxford OX4 1JE, UK

www.myriadeditions.com

This edition published in 2017

1 3 5 7 9 10 8 6 4 2



A CIP catalogue record for this book is available from the British Library.

ISBN: 978-0-9935633-2-4
E-ISBN: 978-0-9935633-3-1

Printed in Poland on paper sourced from sustainable forests.
www.lfbookservices.co.uk

# CONTENTS:

# INTRODUCTION:

**We live in a time when science is under attack. When experts in their fields are openly derided by politicians. Facts are seen as just a matter of opinion. Evidence can be ignored because gut instinct counts for more. The internet and the older news medias are awash with disinformation or outright lies. How are we to pick our way through this sea of information to find the truth?**

*Graphic Science* is about the development of science over a two-hundred-year period, focusing on the lives of seven individuals caught up in the social and political events of their eras. It is a book about war, revolution, gender, race, class, exploration, migration, poverty, sickness, genius and madness. I picked these people not only because the science they engaged in was interesting, but their lives were too.

One theme above others did emerge out of the writing/drawing of this book. It struck me forcibly while researching these seven people that the history of science in Europe and the United States has largely been the history of white, male scientists, with barely a look-in for anyone of another skin colour or gender until the mid-twentieth century. Mary Anning and Washington Carver stand out because they are exceptions. They faced difficulties and barriers that the white, male, more monied scientists, like Antoine Lavoisier and Alfred Wegener, didn't have to. The latter were free to explore and develop their interests as they wished.

How many brilliant women or people of colour have languished in the fields and factories over the centuries, their way to education barred? What developments has humanity as a whole missed out on because of this? Has the pace of scientific progress been slowed from what it could have been had sections of society not been allowed to advance because of gender, race or class? We can never know the answer to these questions for sure, but almost certainly there have been negative consequences for humanity. The education of the majority benefits society as a whole. Education only for those who can afford it is ultimately short-sighted and detrimental to us all. Not just because we as a society lose out in scientific progress, but because poorly educated people are easier to deceive. Knowledge arms us against those who would manipulate us.

The seven people featured in this book were not always right, but they valued the scientific process enough that they were able to make discoveries and open new fields of research. They did not look at the information presented to them and simply believe it. They questioned and looked for evidence. As humans we're bad at accepting truths that run counter to our thinking. Facts rarely change our minds. It can be painful to even read about evidence that undermines our cosy beliefs. Yet it's important that we not only listen to differing viewpoints but accept them when it's clear they have validity. Not only that, but we should question closely even the arguments we agree with, and not just give them a pass because they are comforting. Only in this way can we hope to sort out the truth from the lies. Science itself progresses using methods of critical thinking which we would all do well to learn.

Be a scientist in your own life. Change things the way these seven people did. They were not superhuman. They struggled much as we do. Yet they have transcended their lives and given much to the world. I've tried in these pages to strip away the myths that have built up over time to show the human beings beneath. They speak with their own words wherever possible. So I present to you: Antoine Lavoisier, Mary Anning, George Washington Carver, Nikola Tesla, Alfred Wegener, Jocelyn Bell Burnell and Fred Hoyle. Now read on.

Darryl Cunningham

ANTOINE
LAVOISIER

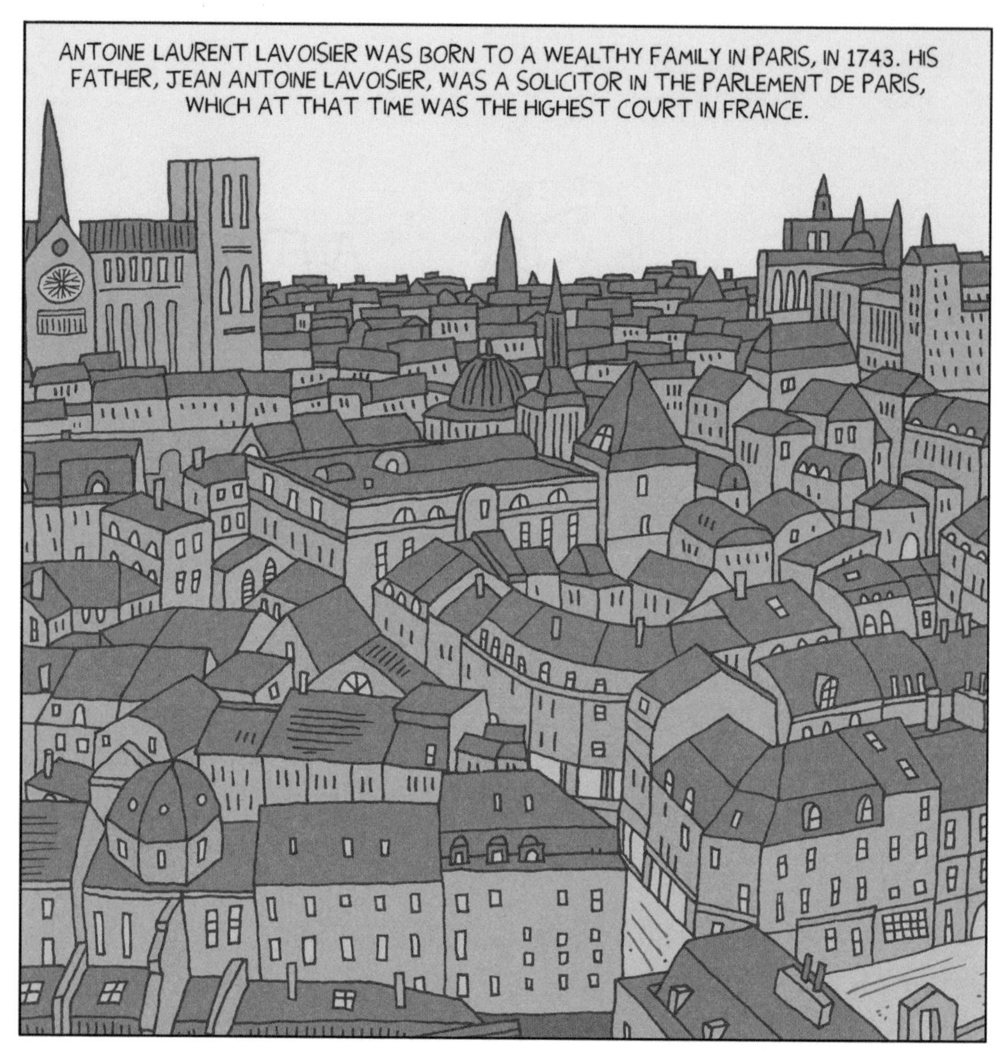
ANTOINE LAURENT LAVOISIER WAS BORN TO A WEALTHY FAMILY IN PARIS, IN 1743. HIS FATHER, JEAN ANTOINE LAVOISIER, WAS A SOLICITOR IN THE PARLEMENT DE PARIS, WHICH AT THAT TIME WAS THE HIGHEST COURT IN FRANCE.

JEAN ANTOINE WANTED HIS ONLY SON TO FOLLOW HIM INTO THE LAW...

BUT THE YOUNG MAN HAD OTHER IDEAS.

THE ACADEMY HAD A FIXED NUMBER OF MEMBERS. VACANCIES AT THE LOWER LEVELS WERE TYPICALLY CREATED BY INTERNAL PROMOTIONS TO THE HIGHER RANKS...

THE GENERAL FARM WAS A FINANCIAL COMPANY THAT COLLECTED TAXES ON BEHALF OF THE ROYAL GOVERNMENT.
GIVE.

THIS INVESTMENT FREED LAVOISIER FROM THE NEED TO SUPPORT HIMSELF BY PRACTISING LAW, AS HE COULD LIVE COMFORTABLY ON INCOME FROM THE FARM.

IN THEORY HE COULD DEDICATE HIS WORKING HOURS ENTIRELY TO SCIENCE.

THIS DIDN'T HAPPEN. INSTEAD HE BECAME VERY INVOLVED WITH PUBLIC SERVICE WORK...

AND THIS OFTEN OVERLAPPED WITH HIS SCIENTIFIC INTERESTS AND RESEARCH.

HE BEGAN AS A REGIONAL INSPECTOR FOR THE FARM'S TOBACCO COMMISSION.
SEIZE THOSE GOODS!

IN HIS WORK FIGHTING THE TRADE IN UNTAXED TOBACCO, LAVOISIER REPORTED TO HIS SUPERIOR, JACQUES-ALEXIS PAULZE DE CHASTEIGNOLLES.

PAULZE, A SENIOR PARTNER IN THE FARM WAS ALSO, LIKE LAVOISIER'S FATHER, A LAWYER IN THE PARLEMENT DE PARIS.

IN 1770, AFTER HIS WIFE'S DEATH, PAULZE BROUGHT HIS DAUGHTER BACK FROM THE CONVENT WHERE SHE'D BEEN EDUCATED, TO SERVE AS HIS HOSTESS.

MARIE-ANNE PIERRETTE PAULZE WAS ONLY THIRTEEN YEARS OLD, BUT WAS ALREADY BESIEGED BY MARRIAGE PROPOSALS.

SHE'D COME TO THE ATTENTION OF AN IMPOVERISHED FIFTY-YEAR-OLD NOBLEMAN – THE COMTE D'AMERVAL.
HEH HEH!

THE MAN IS A FOOL, AN UNFEELING RUSTIC AND AN OGRE.

I'M NOT GOING TO FORCE YOU INTO THIS MARRIAGE IF YOU HAVE SUCH AN AVERSION TO THE MAN.

BUT YOU SHOULD KNOW THAT I'M UNDER POLITICAL PRESSURE TO MAKE THIS MATCH. I COULD EASILY LOSE MY POSITION IN THE GENERAL FARM.

FORTUNATELY I HAVE POWERFUL FRIENDS WHO WILL HELP ME AVERT THIS OUTCOME.

STILL, I SHOULD FIND YOU ANOTHER MATCH, SO THAT WE CAN AVOID SUCH A SITUATION IN THE FUTURE.

HMM! I WONDER WHO WOULD BE SUITABLE?

LAVOISIER.

THE COUPLE WERE MARRIED ON 16TH DECEMBER 1771. THE BRIDE WAS FOURTEEN, THE GROOM EXACTLY TWICE HER AGE.

LIKE MOST 18TH-CENTURY MARRIAGES, THIS UNION WAS A BUSINESS DEAL FOREMOST AND A ROMANCE SECOND.

YET, DESPITE THIS, THE MARRIAGE DID APPEAR TO WORK. MARIE-ANNE TOOK MORE THAN A PASSING INTEREST IN HER HUSBAND'S SCIENTIFIC PURSUITS.

SHE KEPT A METICULOUS RECORD OF ANY EXPERIMENTS AND EDITED HIS REPORTS. HER KNOWLEDGE OF ENGLISH AND LATIN ENABLED HER TO TRANSLATE WORKS BY JOSEPH PRIESTLEY, HENRY CAVENDISH AND OTHERS...

SO THAT LAVOISIER COULD KEEP ABREAST OF CURRENT DEVELOPMENTS IN CHEMISTRY.

PERHAPS HER MOST IMPORTANT TRANSLATION WAS OF RICHARD KIRWAN'S 'ESSAY ON PHLOGISTON AND THE CONSTITUTION OF ACIDS'...

WHICH SHE BOTH TRANSLATED AND CRITIQUED, ADDING FOOTNOTES AND POINTING OUT ERRORS MADE IN THE CHEMISTRY THROUGHOUT THE PAPER.

MARIE-ANNE TOOK COURSES IN CHEMISTRY SO THAT SHE WAS MORE ABLE TO ASSIST HER HUSBAND IN THE LABORATORY.

SHE ALSO TOOK LESSONS IN DRAWING AND PAINTING FROM THE FRENCH NEO-CLASSICAL PAINTER, JACQUES LOUIS DAVID.

THESE SKILLS ENABLED HER TO DO ALL OF THE ILLUSTRATIONS OF LAVOISIER'S EQUIPMENT AND EXPERIMENTS SEEN IN HIS PUBLISHED WORK.

WAS MARIE-ANNE A SCIENTIST? IT COULD BE ARGUED THAT SHE DID NO MORE THAN THE BASIC WORK AN UNDERGRADUATE WOULD DO, AND PERHAPS THAT'S TRUE.

BUT WE THINK OF UNDERGRADUATES AS SCIENTISTS, SO WHY NOT MARIE-ANNE?

APART FROM THE SCIENTIFIC SKETCHES, LITTLE OF HER ARTWORK SURVIVES...

OTHER THAN A PORTRAIT OF BENJAMIN FRANKLIN, WHICH SHE PRESENTED TO HIM AS A GIFT IN 1778.

IT IS ALLOWED BY THOSE WHO HAVE SEEN IT TO HAVE GREAT MERIT AS A PICTURE IN EVERY RESPECT...

BUT WHAT PARTICULARLY ENDEARS IT TO ME IS THE HAND THAT DREW IT.

AFTER THIRTEEN YEARS OF MARRIAGE, MARIE-ANNE BEGAN AN AFFAIR WITH ONE OF HER HUSBAND'S COLLEAGUES, PIERRE SAMUEL DU PONT DE NEMOURS.

THE RELATIONSHIP WAS SO DISCREET THAT NO ONE KNEW OF IT UNTIL AFTER LAVOISIER'S DEATH.

IN 1775 LAVOISIER WAS APPOINTED A COMMISSIONER OF THE ROYAL GUNPOWDER AND SALTPETRE ADMINISTRATION.

HE AND MARIE-ANNE THEN MOVED INTO A NEW BUILDING CREATED FOR THE ADMINISTRATION AT THE PARIS ARSENAL, NEAR THE BASTILLE PRISON.

HERE LAVOISIER BUILT HIMSELF A STATE-OF-THE-ART LABORATORY.
EXCELLENT.

UNDER HIS DIRECTION, GUNPOWDER RESEARCH BECAME A MILITARY SCIENCE PROJECT.

HIS SERIES OF PROGRAMMES WAS DESIGNED TO FIND THE BEST METHODS OF EXTRACTING SALTPETRE AND CREATING ARTIFICIAL SALTPETRE...

AND TO ACHIEVE A MORE PRECISE UNDERSTANDING OF THE CHEMISTRY OF GUNPOWDER.

THESE PROGRAMMES WERE SO SUCCESSFUL THAT BY 1776 THE COUNTRY HAD ENOUGH SURPLUS GUNPOWDER TO OFFER IT TO THE AMERICAN REVOLUTIONARIES.

REVOLUTION WAS IN THE AIR. FRANCE TOO HAD ENTERED A TURBULENT ERA.

LAVOISIER HAD ALREADY DRAWN THE ANGER OF THE RADICAL JOURNALIST AND POLITICIAN JEAN-PAUL MARAT.

BEFORE EVERYTHING ELSE, MARAT THOUGHT OF HIMSELF AS A SCIENTIST. HIS FIRST AND HIS GREATEST AMBITION WAS TO BE HAILED AS ANOTHER NEWTON.

I HAVE WRITTEN TWELVE BOOKS ON SUBJECTS AS DIVERSE AS OPTICS, PHILOSOPHY AND ELECTRICITY...

YET THE ACADEMY OF SCIENCES REFUSES TO TAKE MY THEORIES SERIOUSLY.

MARAT FOCUSED HIS ANGER SPECIFICALLY ON LAVOISIER, WHO AS THE ACADEMY'S LEADING THEORIST ON COMBUSTION...

IN HIS BOOK *RESEARCH ON FLAMES*, PUBLISHED IN 1780, MARAT CLAIMED THAT FIRE WENT OUT IN AN ENCLOSED SPACE...

BECAUSE THE AIR, UNABLE TO ESCAPE, COMPRESSED AND SMOTHERED THE FLAME...

A HYPOTHESIS THAT WOULD HAVE MADE LITTLE SENSE TO THE LOGICALLY MINDED LAVOISIER.
?

THE MYSTERY AS TO WHY FIRE IS EXTINGUISHED IN SUCH A SITUATION WOULD EVENTUALLY BE EXPLAINED BY LAVOISIER HIMSELF.

IT WAS NOT JUST BENJAMIN FRANKLIN WHO FREQUENTED MARIE-ANNE'S REGULAR SALONS.

THE MOST ILLUSTRIOUS MEMBERS OF THE INTERNATIONAL SCIENTIFIC COMMUNITY VISITED THE LAVOISIERS.

A CLERGYMAN, NATURAL PHILOSOPHER, CHEMIST, EDUCATOR AND POLITICAL THEORIST, PRIESTLEY'S FAME SPRANG FROM HIS DISCOVERY OF SEVERAL UNKNOWN GASES FOUND IN THE AIR.

HE TOLD LAVOISIER OF THE PARTICULAR TYPE OF AIR HE'D FOUND WHEN HEATING UP RED CALX OF MERCURY (NOW KNOWN AS MERCURY OXIDE).

THIS AIR HAD A STRANGE PROPERTY. ON A WHIM I DECIDED TO PLACE A LIGHTED CANDLE IN A JAR WITH THE GAS.

I EXPECTED IT TO BE EXTINGUISHED, BUT INSTEAD IT FLARED BRIGHTLY. I THEN INSERTED A RED-HOT WIRE, WHICH GLOWED AS WHITE AS THE SUN..

FOR CENTURIES IT HAD BEEN BELIEVED THAT AIR WAS ONE OF THE FOUR BASIC ELEMENTS THAT MADE UP THE WORLD, ALONG WITH EARTH, FIRE AND WATER.

YET PRIESTLEY HAD SHOWN THAT THE AIR, FAR FROM BEING A BASIC ELEMENT, COULD BE SEPARATED INTO DIFFERENT GASES WHICH HAD THEIR OWN DISTINCT PROPERTIES.

IT SEEMED TO LAVOISIER THAT THE KEY TO UNDERSTANDING THE COMPOSITION OF THE ATMOSPHERE LAY IN THE NATURE OF FIRE ITSELF.

IF HE COULD EXPLAIN THE PROCESS OF FIRE, THEN HE MIGHT USHER IN A REVOLUTION IN CHEMISTRY.

IN THE 18TH CENTURY THE PREVAILING THEORY OF COMBUSTION WAS THE PHLOGISTON THEORY.

PHLOGISTON IS THOUGHT TO PERMEATE ALL COMBUSTIBLE MATTER. IT SETS UP A VIOLENT MOTION WITHIN SUBSTANCES IN THE PRESENCE OF HEAT.

THIS MOTION RESULTS IN FLAME, AND, AS LONG AS THE AIR IS NOT KEPT FROM IT, THE MOTION WILL CONTINUE UNTIL ONLY EARTHY ASH REMAINS.

THAT IS THE THEORY. HOWEVER, I DO NOT BELIEVE THAT PHLOGISTON EXISTS.

ON RETURNING TO ENGLAND, JOSEPH PRIESTLEY CONTINUED HIS INVESTIGATIONS INTO AIR FROM MERCURY CALX, AT HIS HOME-BUILT LABORATORY IN LEEDS.

HIS RESULTS SHOWED THAT THIS AIR WAS NOT JUST AN ESPECIALLY PURE FORM OF COMMON AIR, BUT WAS...

FIVE OR SIX TIMES BETTER THAN COMMON AIR, FOR THE PURPOSE OF RESPIRATION, INFLAMMATION, AND EVERY OTHER USE OF COMMON AIR.

I CALL THIS AIR DEPHLOGISATED AIR, AS IT IS COMMON AIR DEPRIVED OF ITS PHLOGISTON.

HERE WE SEE THE ESSENTIAL DIFFERENCE BETWEEN THE TWO SCIENTISTS.

PRIESTLEY HAD AN EYE FOR CURIOUS FACTS AND A KEEN INTEREST IN EXPERIMENTATION.

IN HIS WORK HE HAD LEARNED MUCH ABOUT THE NATURE OF COMMON AIR.

HE HAD OBSERVED, AS HAD MANY OTHERS, HOW A CANDLE BURNING IN AN ENCLOSED SPACE WAS SOON EXTINGUISHED...

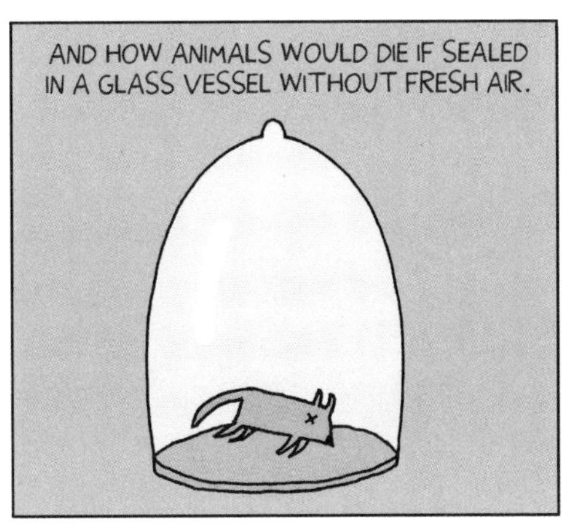
AND HOW ANIMALS WOULD DIE IF SEALED IN A GLASS VESSEL WITHOUT FRESH AIR.

YET, WHEN HE PLACED A MINT PLANT IN A BELL JAR WITH A MOUSE AND A CANDLE, HE SAW THAT THE MOUSE LIVED, AND THE CANDLE CONTINUED TO BURN.

PRIESTLEY CONCLUDED THAT PLANTS RESTORED TO THE AIR WHATEVER IT WAS THAT ANIMALS AND FIRE REMOVED.

YET, AS BRILLIANT AS PRIESTLEY WAS, HE COULD NOT FREE HIMSELF FROM HIS BELIEF IN THE PHLOGISTON THEORY...

AND THIS PREVENTED HIM FROM ACHIEVING A COMPREHENSIVE EXPLANATION OF ALL HE OBSERVED.

HE LACKED LAVOISIER'S TEMPERAMENT FOR DEDUCTIVE REASONING AND WILLINGNESS TO EMBRACE NEW CONCEPTS.

IN HIS OWN INVESTIGATIONS, LAVOISIER HAD FOUND THAT, WHEN HE BURNED PHOSPHORUS, IT GAINED WEIGHT AS IT FORMED PHOSPHORIC ACID.

THE BURNING OF SULPHUR PRODUCED A SIMILAR INCREASE IN WEIGHT.

IF THE PHLOGISTON THEORY WERE CORRECT, THEN THERE SHOULD BE LESS MATERIAL IN THE BURNED REMAINS, NOT MORE.

IT WAS CLEAR TO LAVOISIER THAT SULPHUR, PHOSPHORUS AND OTHER SUBSTANCES WERE ABSORBING SOMETHING IN THE ATMOSPHERE RATHER THAN RELEASING IT.

BUT WHAT WAS THIS SOMETHING?

LAVOISIER EVENTUALLY CONCLUDED THAT THERE WERE TWO COMPONENTS TO THE AIR AROUND US.

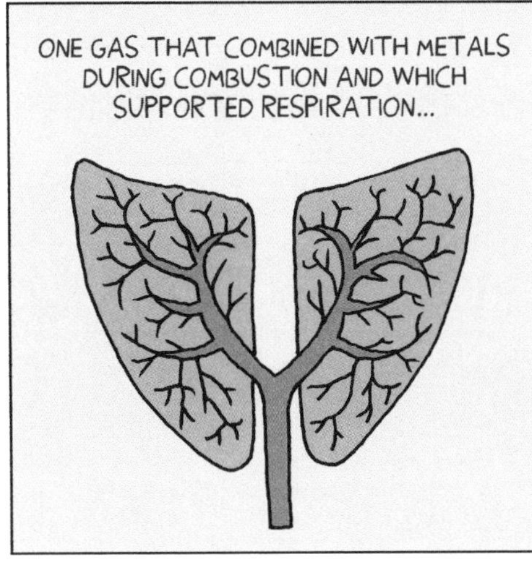

ONE GAS THAT COMBINED WITH METALS DURING COMBUSTION AND WHICH SUPPORTED RESPIRATION...

AND A SECOND GAS THAT DIDN'T SUPPORT EITHER COMBUSTION OR RESPIRATION.

LAVOISIER INCORRECTLY BELIEVED THAT THE BREATHABLE PORTION OF THE AIR WAS NECESSARY TO FORM ALL ACIDS, WHICH WAS WHY HE NAMED IT OXYGEN, FROM THE TWO GREEK WORDS MEANING ACID GENERATOR.

LAVOISIER AND PRIESTLEY WERE NOT THE ONLY PEOPLE TO HAVE A CLAIM ON THE DISCOVERY OF OXYGEN. THE SWISS CHEMIST CARL WILHELM SCHEELE HAD ISOLATED OXYGEN IN 1773 OR EARLIER.

OF THE THREE MEN, PRIESTLEY HAS PROMINENCE, BECAUSE HE PUBLISHED FIRST.

SCHEELE WORKED IN A STOCKHOLM PHARMACY WHICH WAS NOWHERE NEAR AS EXPENSIVELY EQUIPPED AS LAVOISIER'S LABORATORY.

HIS IMPROVISED EQUIPMENT LIMITED WHAT HE COULD ACHIEVE. NEVERTHELESS, SCHEELE WAS A COUPLE OF YEARS AHEAD OF HIS TWO RIVALS.

HOWEVER, LIKE PRIESTLEY, SCHEELE MADE THE MISTAKE OF ONLY BEING ABLE TO SEE OXYGEN'S PROPERTIES IN TERMS OF THE PHLOGISTON THEORY.
I CALL THIS NEW GAS 'FIRE AIR'.

ONLY LAVOISIER WAS PREPARED TO ENTIRELY DISPENSE WITH PHLOGISTON AS A CONCEPT AND START AFRESH.

IN 1783 LAVOISIER READ TO THE ACADEMY HIS PAPER ENTITLED 'REFLECTIONS ON PHLOGISTON' – A FULL-SCALE ATTACK ON THE PHLOGISTON THEORY.

THAT YEAR HE ALSO BEGAN A SERIES OF EXPERIMENTS ON THE COMPOSITION OF WATER.

MANY INVESTIGATORS OF THE ERA HAD EXPERIMENTED WITH THE GAS THAT THE ENGLISH CHEMIST HENRY CAVENDISH CALLED 'INFLAMMABLE AIR', AND WHICH LAVOISIER CALLED HYDROGEN...

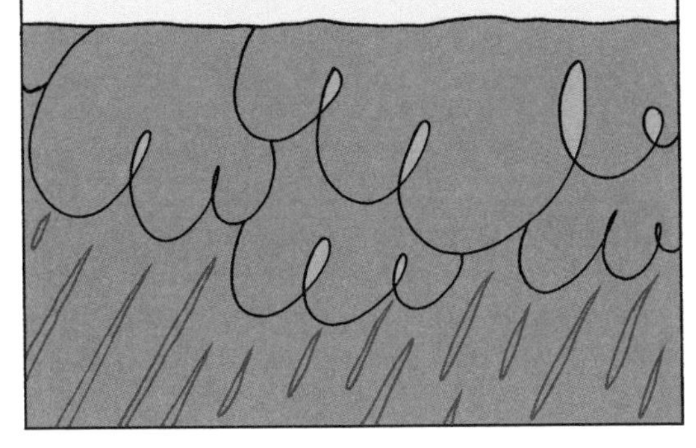

COMBINING IT WITH OXYGEN BY ELECTRICALLY SPARKING MIXTURES OF THE TWO GASES.

ALL RESEARCHERS SAW THAT THESE EXPERIMENTS PRODUCED WATER, BUT THEY ALL EXPLAINED THIS RESULT IN TERMS OF THE PHLOGISTON THEORY.

WITH THE MATHEMATICIAN PIERRE-SIMON LAPLACE, LAVOISIER SYNTHESISED WATER BY BURNING JETS OF HYDROGEN AND OXYGEN IN A BELL JAR OVER MERCURY.

THE RESULTS SUPPORTED THE CONTENTION THAT WATER WAS NOT A BASIC ELEMENT OF MATTER, AS HAD BEEN THOUGHT FOR 2,000 YEARS...

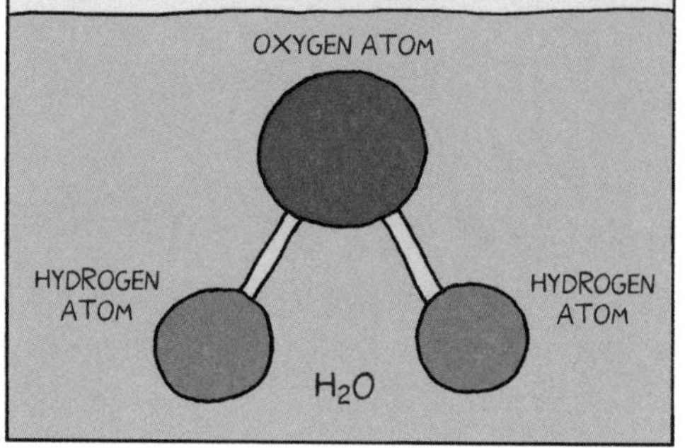
BUT A COMPOUND OF TWO GASES, HYDROGEN AND OXYGEN. IT'S NOW KNOWN THAT THE WATER MOLECULE IS FORMED FROM TWO HYDROGEN ATOMS AND ONE OXYGEN ATOM.
OXYGEN ATOM
HYDROGEN ATOM
HYDROGEN ATOM
$H_2O$

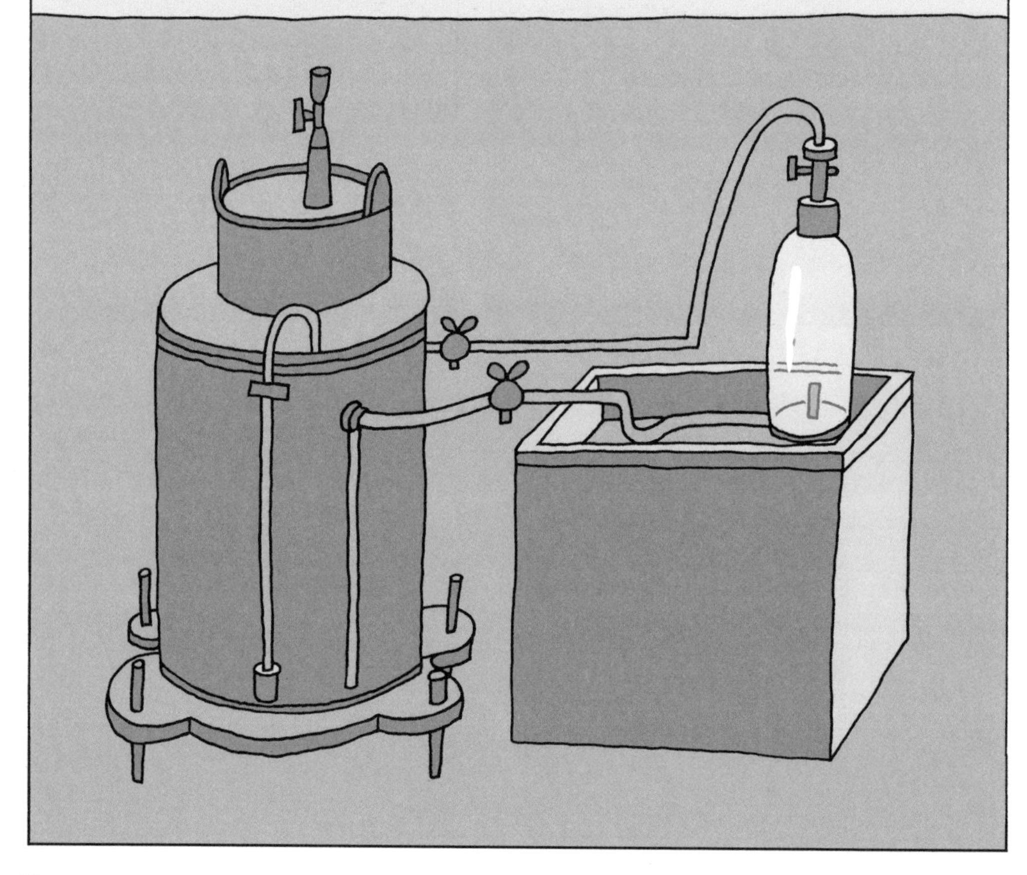
TWO YEARS LATER, WITH THE HELP OF PHYSICIST AND ENGINEER JEAN BAPTISTE MEUSNIER, LAVOISIER RETURNED TO EXPERIMENTS ON THE COMPOSITION OF WATER. THE TWO MEN DESIGNED IMPRESSIVE NEW INSTRUMENTS THAT WERE MINUTELY ACCURATE, EFFICIENT AND BEAUTIFULLY FINISHED. THIS EQUIPMENT WAS EXPENSIVE. JUST ONE OF THE TWO GASOMETERS COST THE EQUIVALENT OF 16,000 POUNDS IN MODERN MONEY.

THAT LAVOISIER'S WEALTH ENABLED HIM TO CREATE SUCH PRECISION INSTRUMENTS GAVE HIM AN ADVANTAGE HIS COMPETITORS DIDNT HAVE.

THIS, TOGETHER WITH HIS REMARKABLE ANALYTICAL MIND, ALLOWED HIM TO TURN CHEMISTRY INTO AN EXACT SCIENCE – ONE BASED ON CAREFUL MEASUREMENT.

THE ACCURATE WEIGHING OF MATERIALS MEANT LAVOISIER WAS ABLE TO PROVE THE LAW OF CONSERVATION OF MASS, WHICH HAD BEEN LONG SUSPECTED, BUT NEVER PROVEN.

THIS LAW STATES THAT MATTER CAN BE CHANGED INTO ONE FORM OR ANOTHER...

MIXTURES CAN BE SEPARATED OR MADE, AND PURE SUBSTANCES CAN BE DECOMPOSED...

BUT THE TOTAL MASS WILL ALWAYS REMAIN THE SAME.

BUT THE NUMBER OF ATOMS BEFORE AND AFTER THE REACTION WILL ALWAYS BE THE SAME.

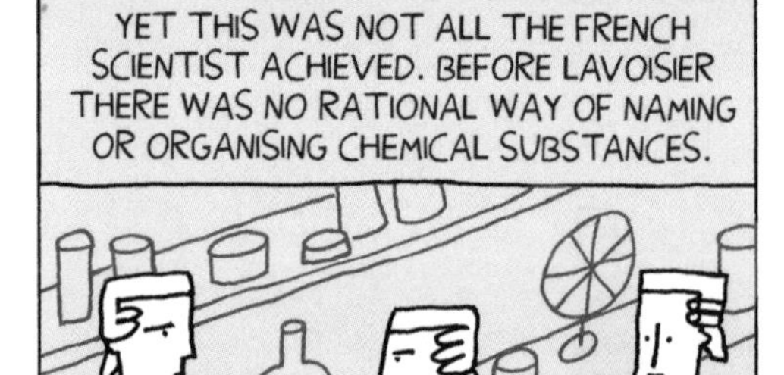

THE CLASSICAL ELEMENTS OF EARTH, AIR, FIRE AND WATER WERE DISCARDED, AND INSTEAD SOME 55 ELEMENTS WHICH COULD NOT BE BROKEN DOWN INTO SIMPLER SUBSTANCES WERE LISTED. CURIOUSLY, AND IN ERROR, THIS LIST ALSO INCLUDED LIGHT, AND A FICTITIOUS SUBSTANCE CALLED CALORIC.

IN 1789 LAVOISIER PUBLISHED HIS *ELEMENTARY TREATISE ON CHEMISTRY.* THIS WORK REPRESENTED THE SYNTHESIS OF HIS CONTRIBUTION TO SCIENCE.

IT IS CONSIDERED TO BE THE FIRST MODERN TEXTBOOK ON THE SUBJECT. THE CORE OF THE WORK WAS THE OXYGEN THEORY.
TRAITÉ ÉLÉMENTAIRE DE CHIMIE
ANTOINE LAVOISIER

THE BOOK PRESENTED A UNIFIED VIEW OF THE NEW THEORIES OF CHEMISTRY. IT CONTAINED A CLEAR STATEMENT OF THE LAW OF CONSERVATION OF MASS.

IT DENIED THE EXISTENCE OF PHLOGISTON AND MADE CLEAR THE DIFFERENCE BETWEEN AN ELEMENT (A PURE SUBSTANCE MADE OF THE SAME TYPE OF ATOM) AND A COMPOUND (A SUBSTANCE CONTAINING DIFFERENT ELEMENTS).

EVEN THOUGH LAVOISIER HAD ESSENTIALLY DRAWN FROM THE WORK OF JOSEPH PRIESTLEY, HENRY CAVENDISH, JOSEPH BLACK...

CARL WILHELM SCHEELE AND OTHERS, IT WAS STILL A CONSIDERABLE ACHIEVEMENT.

AS A COMMISSIONER OF THE GUNPOWDER ADMINISTRATION, THE SCIENTIST HAD SPENT MUCH OF THAT DAY BEFORE THE NATIONAL ASSEMBLY, TRYING TO JUSTIFY THE TRANSFER OF THE GUNPOWDER.

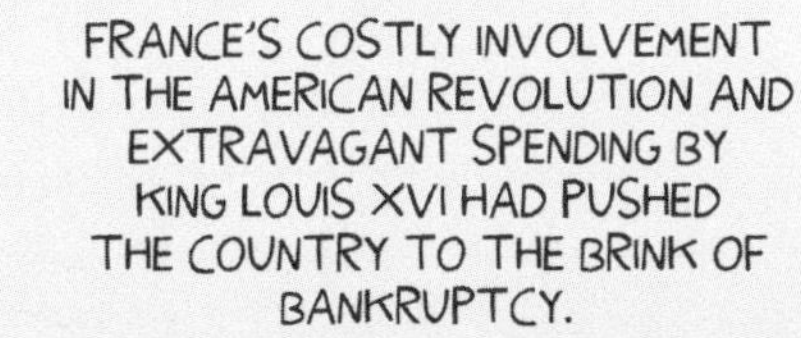

TWO DECADES OF BAD CEREAL HARVESTS, DROUGHT, CATTLE DISEASES AND SKYROCKETING BREAD PRICES HAD KINDLED UNREST AMONG THE POOR.

MANY EXPRESSED THEIR DESPERATION AND RESENTMENT TOWARDS A MONARCHY THAT IMPOSED HEAVY TAXES BUT GAVE LITTLE RELIEF...

DURING THE REVOLUTION, FRENCH CITIZENS RAZED AND REDESIGNED THEIR COUNTRY'S POLITICAL LANDSCAPE, UPROOTING CENTURIES-OLD INSTITUTIONS.

BY 1793 BOTH MONARCHY AND THE FEUDAL SYSTEM WERE GONE.

REGIME CHANGE DID NOT COME WITHOUT BLOODSHED. 40,000 PEOPLE WERE EXECUTED OR DIED IN PRISON DURING THE YEARS OF 'THE TERROR'.

SOME PEOPLE DIED FOR THEIR POLITICAL OPINIONS, BUT MANY FOR LITTLE REASON, BEYOND MERE SUSPICION OR BECAUSE OTHERS WANTED THEM OUT OF THE WAY.

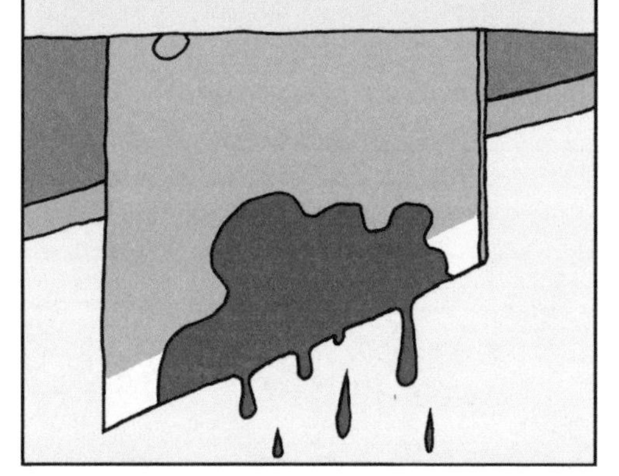

AS THE PACE OF THE REVOLUTION QUICKENED, LAVOISIER DID WHAT HE COULD TO DETACH HIMSELF FROM POLITICS.

HE PUBLISHED AN OPEN LETTER IN *LE MONITEUR UNIVERSEL*, STATING THAT HE WAS DECLINING THE SALARIES FOR FOUR OF THE FIVE PUBLIC OFFICES HE HELD...

THE EXCEPTION BEING THE GUNPOWDER ADMINISTRATION WHERE HE WANTED TO RETAIN THE POST IN ORDER TO KEEP HIS LABORATORY AT THE ARSENAL.

WAS HE NOT THE MOST PROMINENT SCIENTIST IN ALL FRANCE? HAD HE NOT BEEN MORE ENERGETIC, MORE DILIGENT AND MORE SUCCESSFUL THAN ANY OF HIS RIVALS?

NEVERTHELESS, IN 1792, HE WAS FORCED TO RESIGN FROM HIS POST AT THE GUNPOWDER COMMISSION...

SIGH!

IN JULY 1793, MARAT WAS MURDERED – STABBED THROUGH THE HEART BY A YOUNG WOMAN CALLED CHARLOTTE CORDAY, WHILE HE LAY IN THE BATH.

THE DEATH OF HIS ENEMY DID NOT END LAVOISIER'S TROUBLES. THE HOSTILITY TOWARDS HIM FROM THE NEW REGIME ONLY INCREASED.

THE BLOW CAME ON 24TH NOVEMBER 1794, WHEN THE ORDER WENT OUT TO ARREST ALL FORMER MEMBERS OF THE GENERAL FARM.

LAVOISIER TOOK REFUGE IN THE ACADEMY OF SCIENCES' BUILDING, WHICH HAD BEEN DESERTED SINCE THE ACADEMY'S ABOLITION THE PREVIOUS YEAR.

FOR FOUR DAYS HE HID IN THE EMPTY STOREROOMS AND LECTURE HALLS, WRITING LETTERS IN DEFENCE OF HIS CASE.

FINALLY, WHEN HE HEARD THAT HIS FATHER-IN-LAW, JACQUES PAULZE, HAD BEEN ARRESTED, LAVOISIER GAVE HIMSELF UP.

ANTOINE LAVOISIER WENT TO THE GUILLOTINE ON 8TH MAY 1794, ALONG WITH JACQUES PAULZE AND 27 CO-DEFENDANTS.

THE NOTED MATHEMATICIAN LOUIS LAGRANGE SAID ABOUT THE EXECUTION...
IT TOOK THEM ONLY AN INSTANT TO CUT OFF THAT HEAD, AND ONE HUNDRED YEARS MIGHT NOT SUFFICE TO REPRODUCE ITS LIKE.

ALTHOUGH MARIE-ANNE WAS THROWN INTO BANKRUPTCY AFTER LAVOISIER'S DEATH, HER PROPERTY AND MONEY WERE REINSTATED EIGHTEEN MONTHS LATER WHEN HER HUSBAND WAS EXONERATED.

A WEALTHY WIDOW, MARIE-ANNE MADE A SHORT-LIVED MARRIAGE TO ANOTHER SCIENTIST, THE AMERICAN BENJAMIN THOMPSON, BETTER KNOWN AS COUNT RUMFORD.

IN A MALE-DOMINATED SOCIETY, MARIE-ANNE'S ONLY ACCESS TO SCIENCE WOULD HAVE BEEN THROUGH ANOTHER MAN. WHEN THIS STRATEGY FAILED, THAT DOOR WAS FIRMLY SHUT.

THE MARRIAGE WAS NOT HELPED BY THE FACT THAT MARIE-ANNE INSISTED ON KEEPING HER FIRST HUSBAND'S NAME. SHE REMAINED DEVOTED TO LAVOISIER'S MEMORY FOR THE REST OF HER LIFE.

SHE WAS ABLE TO RECOVER NEARLY ALL LAVOISIER'S NOTEBOOKS AND CHEMICAL APPARATUS. SHE EDITED HIS UNFINISHED BOOK, *MEMOIRS OF CHEMISTRY*, WHICH WAS FINALLY PUBLISHED IN 1805.

IN HER LATTER YEARS SHE PRESIDED OVER A THRICE-WEEKLY GATHERING OF ARTISTS, SCIENTISTS AND POETS, IN WHAT WAS DESCRIBED AS THE LAST EXAMPLE OF A FRENCH 18TH-CENTURY SALON.

NEVER CONSIDERED A SCIENTIST IN HER OWN RIGHT, SHE WAS AS MUCH OF A SCIENTIST AS THE SOCIETY OF THE TIME WOULD ALLOW. SHE DIED IN 1832 AT THE AGE OF 78.
END

MARY
ANNING

MARY ANNING WAS BORN IN LYME REGIS, DORSET, ENGLAND, IN 1799.

HER FATHER, RICHARD ANNING, WAS A SKILLED CARPENTER WHO STRUGGLED TO MAKE A LIVING.

HE WOULD SUPPLEMENT HIS MEAGRE INCOME BY MINING THE COASTAL CLIFFS OF FOSSILS AND SELLING THEM TO TOURISTS.

WHEN MARY ANNING WAS BORN, BRITAIN HAD BEEN AT WAR WITH REVOLUTIONARY FRANCE FOR SIX YEARS. SHE WOULD BE SIXTEEN BEFORE IT WAS OVER.
LYME REGIS
PARIS

THIS WAR HAD TWO EFFECTS – IT TURNED LYME REGIS INTO A TOURIST RESORT FOR WELL-OFF GENTRY WHO WERE UNABLE TO VISIT THE CONTINENT...

AND IT CREATED FOOD SHORTAGES AND HIGH PRICES, ESPECIALLY OF BREAD, DUE TO THE LACK OF EUROPEAN CORN ON THE MARKET.

LIFE BECAME HARDER FOR THOSE WHO WERE ALREADY POOR.

THE ANNINGS WERE FORTUNATE IN THAT FOSSILS WERE CLOSE AT HAND TO PROVIDE THEM WITH EXTRA INCOME.

RICHARD AND HIS WIFE, MOLLY, HAD AT LEAST NINE CHILDREN, PERHAPS TEN, BUT ONLY TWO WERE TO SURVIVE INTO ADULTHOOD.

MARY ANNING HERSELF NARROWLY ESCAPED DEATH WHEN SHE WAS AN INFANT.

BABY MARY WAS BEING HELD BY A MRS ELIZABETH HASKINS IN THE COMPANY OF TWO TEENAGE GIRLS...

WHEN THEY WERE STRUCK BY LIGHTNING.

OH, MY LORD!

MRS HASKINS AND THE TWO GIRLS WERE KILLED...

BUT MARY WAS FOUND MIRACULOUSLY ALIVE.

THE ANNING FAMILY WERE MEMBERS OF A DISSENTING FACTION OF PROTESTANT CHRISTIANS WHO HAD SPLIT FROM THE ESTABLISHED CHURCH OF ENGLAND.

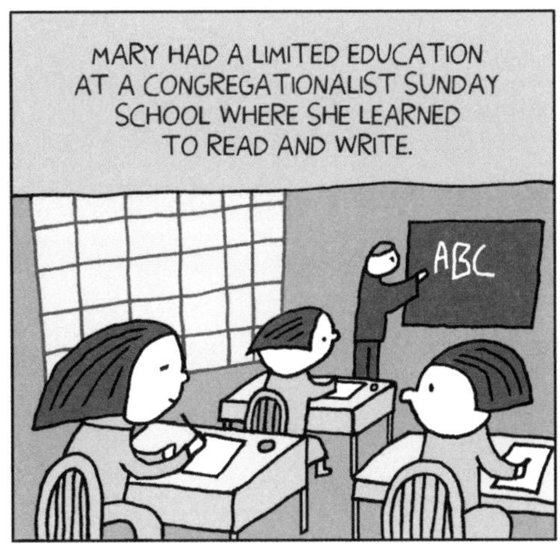
MARY HAD A LIMITED EDUCATION AT A CONGREGATIONALIST SUNDAY SCHOOL WHERE SHE LEARNED TO READ AND WRITE.
ABC

RICHARD ANNING'S FELLOW CONGREGATIONALISTS WOULD HARDLY HAVE APPROVED OF HIS WORKING ON A SUNDAY.

YET, WHEN HIS CARPENTRY WORK WAS IN SHORT SUPPLY, HE WOULD HAVE HAD LITTLE CHOICE BUT TO SCOUR THE BEACHES FOR FOSSILS, EVEN ON THE DAY OF REST.

EVEN THIS EXTRA INCOME WAS BARELY ENOUGH TO FEND OFF STARVATION. A RIVAL FOSSIL-HUNTER, WHOM RICHARD ANNING KNEW, WAS A MR CRUIKSHANKS (OR CROOKSHANKS).
MORNING!
MORNING!

IN 1802, WHEN THE COAL MERCHANT BUSINESS HE HAD LEFT BEHIND IN LONDON STOPPED SENDING HIM A SMALL ANNUITY...

CROOKSHANKS REALISED THAT THE MONEY HE MADE FROM SELLING FOSSILS WOULD NOT BE ENOUGH TO KEEP HIM FROM THE POORHOUSE.

SO HE JUMPED TO HIS DEATH FROM GUN CLIFF, ONLY A FEW YARDS FROM COCKMOILE SQUARE WHERE THE ANNINGS LIVED.

WHAT WERE FOSSILS? IN THE EARLY 19TH CENTURY PEOPLE DID NOT KNOW.

IT WAS A TIME WHEN MOST HAD ABSOLUTE FAITH IN THE WORD OF THE BIBLE. THE GENERAL VIEW WAS THAT FOSSILS WERE THE REMAINS OF CREATURES...

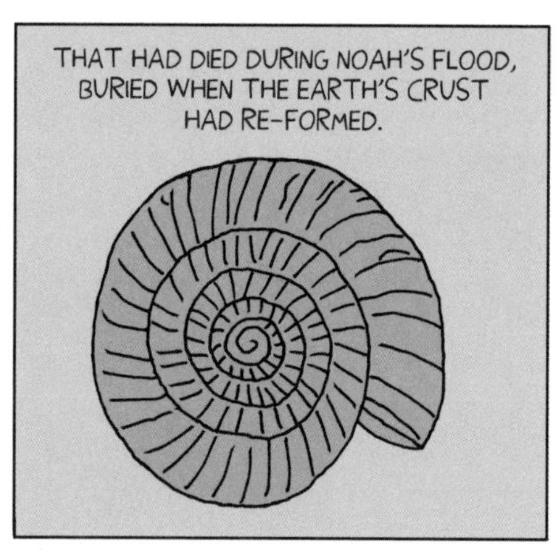
THAT HAD DIED DURING NOAH'S FLOOD, BURIED WHEN THE EARTH'S CRUST HAD RE-FORMED.

IT WAS BELIEVED THAT THE EARTH WAS ONLY A FEW THOUSAND YEARS OLD. PEOPLE DID NOT THINK IN TERMS OF MILLIONS OF YEARS, AND CERTAINLY NOT BILLIONS.

IN 1650 THE ARCHBISHOP OF ARMAGH, JAMES USSHER, CONCLUDED THAT GOD HAD CREATED THE WORLD ON SUNDAY 23RD OCTOBER, 4,000 YEARS BEFORE THE BIRTH OF CHRIST.

HE MADE THESE CALCULATIONS BY ADDING THE LIFE SPANS OF THE DESCENDANTS OF ADAM, COMBINED WITH HIS KNOWLEDGE OF THE HEBREW CALENDAR, AND OTHER BIBLICAL RECORDS.

THIS APPROACH WAS NOT RIDICULED, BUT ACCEPTED AS AN EXCELLENT PIECE OF SCHOLARSHIP.

AS THE 19TH CENTURY PROGRESSED, SUCH RELIGIOUS CERTAINTY WOULD BEGIN TO ERODE.

THE STUDY OF FOSSILS WAS TO BE AT THE FOREFRONT OF THIS NEW THINKING.

FOSSIL DISCOVERIES CHALLENGED THE ACCOUNT OF THE CREATION OF THE EARTH AS DEPICTED IN GENESIS, AND THIS WAS TROUBLING TO MANY PEOPLE.

ONE NIGHT, IN THE WINTER OF 1809/10, RICHARD ANNING, WHILE ON HIS WAY TO NEARBY CHARMOUTH...

FELL DOWN THE SLOPES OF THE TREACHEROUS 'BLACK VEN' CLIFF.

WEAKENED BY HIS INJURIES, HE SUCCUMBED TO TUBERCULOSIS THE FOLLOWING AUTUMN.

HE LEFT TWO CHILDREN AND A HEAVILY PREGNANT WIFE, DESTITUTE.

MOLLY ANNING HAD TO APPLY TO THE PARISH FOR POOR RELIEF. SHE RECEIVED IN RETURN ONLY A SMALL AMOUNT FOR FOOD AND CLOTHING.

MORE THAN EVER THE ANNING FAMILY NEEDED THE MONEY GENERATED BY THE FINDING OF FOSSILS FROM THE BEACH CLIFFS OF LYME REGIS.

LATE IN 1810 OR EARLY 1811, JOSEPH ANNING, MARY'S OLDER BROTHER, MADE A DISCOVERY AMONG THE ROCKS BELOW BLACK VEN.

THE FOSSILISED SKULL OF A LARGE ANIMAL.

JOSEPH WENT TO FETCH MARY AND THE TWO RETURNED TO THE SPOT.
WHERE'S THE REST OF IT, I WONDER?

I RECKON IT'S UNDER ALL THAT RUBBLE AND MUD THAT'S RECENTLY FALLEN FROM THE CLIFF.

WE'LL NEVER MOVE ALL THAT LOT AND IT'LL BE AGES BEFORE THE SEA WASHES IT AWAY.

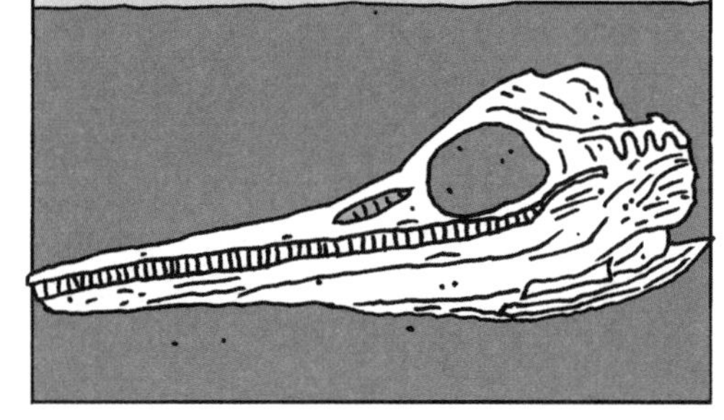
IT WAS ALMOST A YEAR LATER, WHEN MARY WAS TWELVE OR THIRTEEN AND THE SEA HAD DONE ITS WORK REARRANGING THE CLIFFSIDE, THAT SHE FOUND THE REST OF THE CREATURE NOT FAR FROM WHERE THE SKULL HAD BEEN UNEARTHED.

WORKING WITH HER HAMMER, SHE FIRST FOUND A LONG BACKBONE, AND THEN THE OUTLINE OF A FEW RIBS.

WHAT WAS GRADUALLY UNCOVERED WAS A CREATURE UP TO 17 FEET LONG.

THE FAMILY HAD TO HIRE WORKMEN TO HELP CUT THE FOSSIL FROM THE ROCK.

THE COMPLETE SPECIMEN WAS SOLD TO THE LOCAL LORD OF THE MANOR, HENRY HOSTE HENLEY, FOR THE SUM OF 23 POUNDS. ENOUGH TO FEED THE ANNINGS FOR SIX MONTHS.

HENLEY SOLD THE FOSSIL ON TO WILLIAM BULLOCK'S MUSEUM OF NATURAL CURIOSITIES IN PICCADILLY, LONDON...

WHERE IT WAS WRONGLY LABELLED AS A CROCODILE.

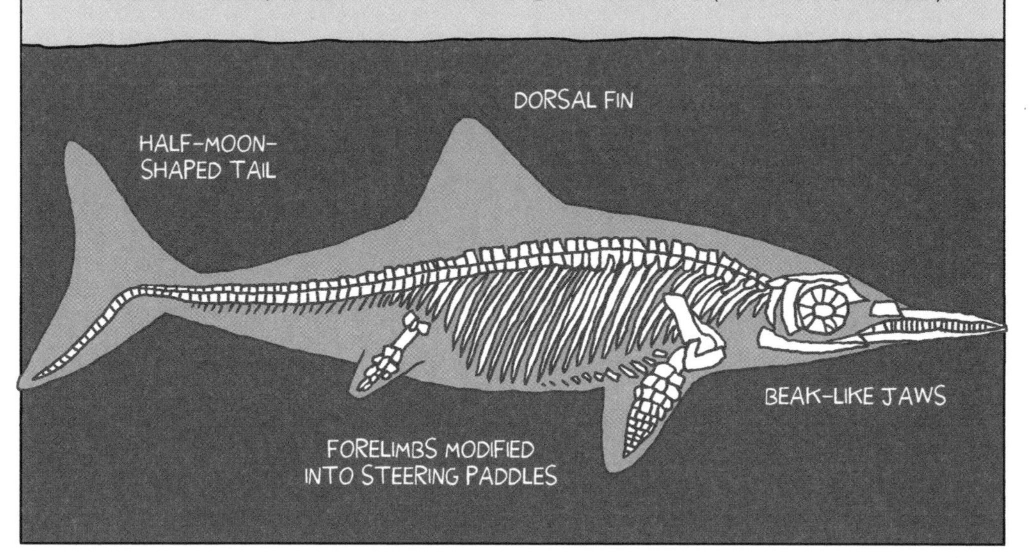
BULLOCK'S ODD COLLECTION - COMPOSED OF ITEMS BROUGHT BACK BY CAPTAIN COOK FROM THE SOUTH SEAS, MEMORABILIA OF NAPOLEON, STUFFED ANIMALS AND EXHIBITS FROM EGYPT AND MEXICO - WAS AUCTIONED OFF IN 1819. MARY'S FOSSIL WAS BOUGHT BY THE BRITISH MUSEUM WHERE AN ASSISTANT CURATOR, CHARLES KONIG, SUGGESTED THE NAME ICHTHYOSAURUS (MEANING FISH LIZARD).
DORSAL FIN
HALF-MOON-SHAPED TAIL
BEAK-LIKE JAWS
FORELIMBS MODIFIED INTO STEERING PADDLES

KONIG HAD NOTICED THE CURIOUS MIXTURE OF FISH AND REPTILE CHARACTERISTICS. IT WAS ALSO CLEAR THAT THE ICHTHYOSAURUS SHARED MANY SIMILARITIES WITH THE MODERN-DAY DOLPHIN.

FOR AN ANIMAL TO BE FOSSILISED, IT HAS TO DIE IN A PLACE PROTECTED FROM SCAVENGERS AND THE ELEMENTS...

AND THIS MEANS BURIAL IN SAND, SOIL OR MUD, USUALLY AT THE BOTTOM OF A RIVER, OR ON THE SEABED.

OVER TIME THE FLESH AND SOFT PARTS ARE BROKEN DOWN BY BACTERIA, AND THE SKELETON IS BURIED DEEPER, AS FURTHER SEDIMENT PILES ON TOP OF THE REMAINS.

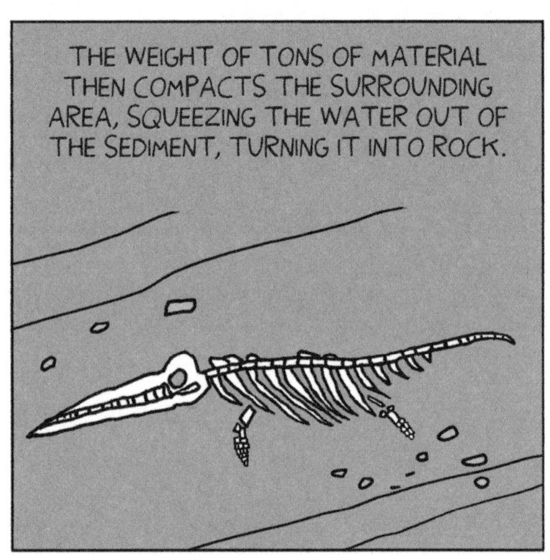
THE WEIGHT OF TONS OF MATERIAL THEN COMPACTS THE SURROUNDING AREA, SQUEEZING THE WATER OUT OF THE SEDIMENT, TURNING IT INTO ROCK.

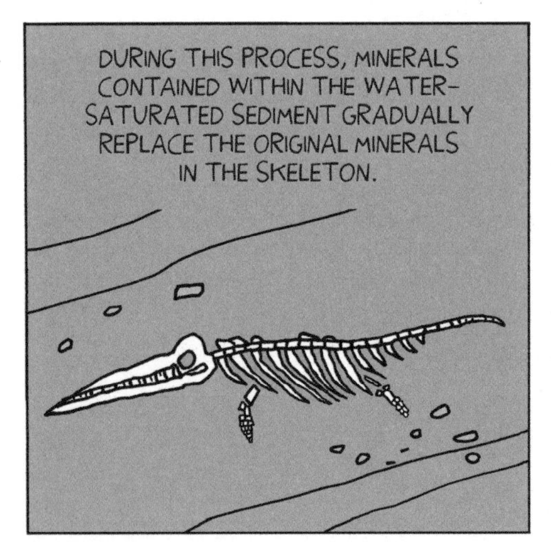
DURING THIS PROCESS, MINERALS CONTAINED WITHIN THE WATER-SATURATED SEDIMENT GRADUALLY REPLACE THE ORIGINAL MINERALS IN THE SKELETON.

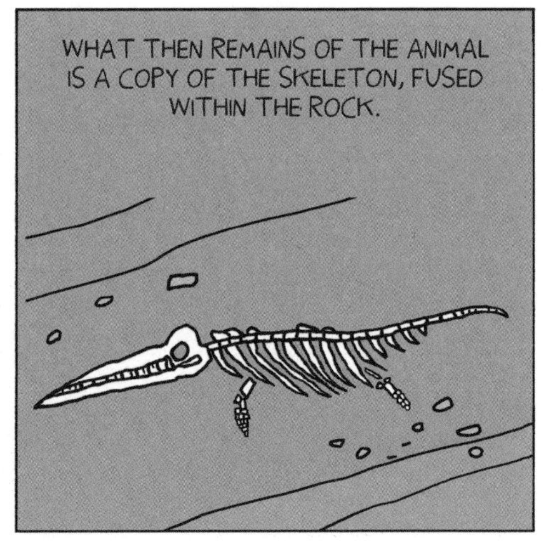
WHAT THEN REMAINS OF THE ANIMAL IS A COPY OF THE SKELETON, FUSED WITHIN THE ROCK.

WHY IS LYME REGIS SO RICH IN FOSSILS? IT'S BECAUSE THE CLIFFS BETWEEN LYME AND CHARMOUTH CONTAIN FOSSILS FROM THE LOWER JURASSIC PERIOD...

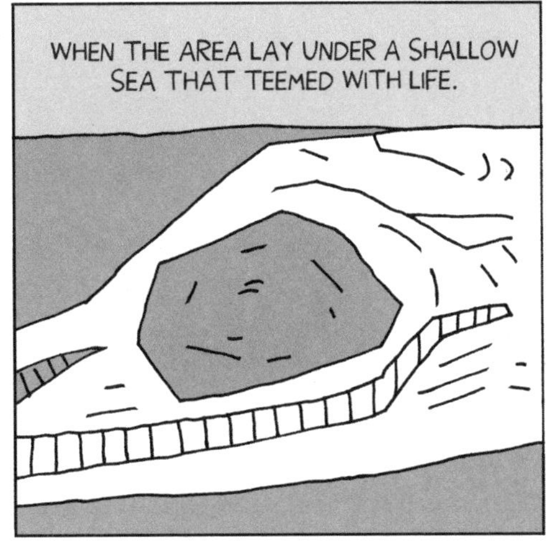
WHEN THE AREA LAY UNDER A SHALLOW SEA THAT TEEMED WITH LIFE.

AFTER MILLIONS OF YEARS THE LAYERS OF PETRIFIED SEDIMENT ARE UNDER CONSTANT ATTACK FROM THE ELEMENTS.

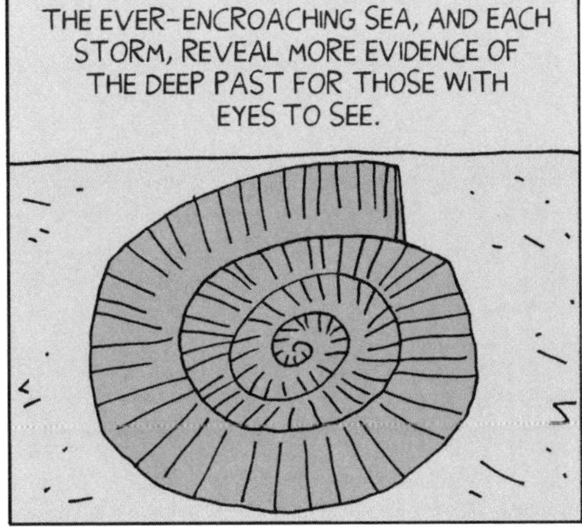
THE EVER-ENCROACHING SEA, AND EACH STORM, REVEAL MORE EVIDENCE OF THE DEEP PAST FOR THOSE WITH EYES TO SEE.

THE ANNING FAMILY'S DISCOVERY OF THE ICHTHYOSAURUS ATTRACTED MANY ENTHUSIASTIC GEOLOGISTS AND FOSSIL-HUNTERS TO THE TOWN...

ALL WEALTHY MEN WHO HAD THE MONEY TO INDULGE THEIR SCIENTIFIC INTERESTS.

MANY OF THE PIONEERS OF THESE NEW SCIENCES WERE CLERGYMEN: MEN WHOSE POSITION IN THE CHURCH GAVE THEM THE FINANCIAL SECURITY TO CONTINUE WITH THEIR STUDIES.

AMONG THOSE WHO HIRED MARY ANNING AS A GUIDE TO SEARCH FOR FOSSILS ALONG THE CLIFFS AT LYME WAS WILLIAM BUCKLAND.

BUCKLAND WAS TO BE THE FIRST PROFESSOR OF GEOLOGY AT OXFORD AND EVENTUALLY A CANON AT CHRIST CHURCH CATHEDRAL...
CAREFUL, SIR. THOSE ROCKS ARE SLIPPERY.

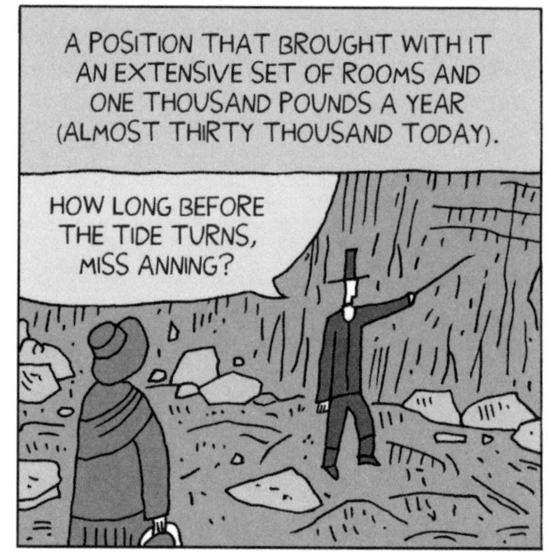
A POSITION THAT BROUGHT WITH IT AN EXTENSIVE SET OF ROOMS AND ONE THOUSAND POUNDS A YEAR (ALMOST THIRTY THOUSAND TODAY).
HOW LONG BEFORE THE TIDE TURNS, MISS ANNING?

FURTHERMORE, NO REAL DUTIES WERE REQUIRED IN THIS POST.
NOT LONG, SIR.

SUCH FREE TIME WOULD ENABLE BUCKLAND TO BECOME PRESIDENT OF THE GEOLOGICAL SOCIETY OF LONDON IN 1824.
WE'D BEST BE MAKING OUR WAY BACK BEFORE THE SEA CUTS US OFF.

SUCH OPPORTUNITIES DID NOT EXIST FOR WOMEN OF MARY ANNING'S CLASS AND EDUCATION.

AT THIS TIME IN BRITAIN, WOMEN COULD NOT VOTE, HOLD PUBLIC OFFICE OR ATTEND UNIVERSITY.

THE GEOLOGICAL SOCIETY OF LONDON DID NOT ALLOW WOMEN TO BECOME MEMBERS OR EVEN ATTEND AS GUESTS.

MARY BECAME A FOSSIL-HUNTER BECAUSE SHE HAD NO OTHER CHOICE. HER BROTHER, JOSEPH, COULD AT LEAST TAKE AN UPHOLSTERY APPRENTICESHIP.

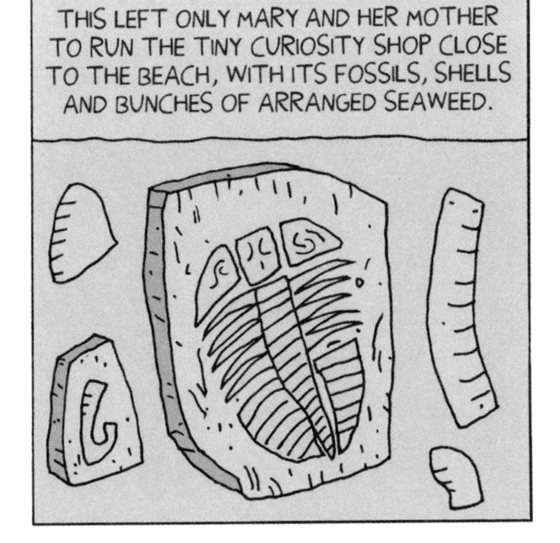
THIS LEFT ONLY MARY AND HER MOTHER TO RUN THE TINY CURIOSITY SHOP CLOSE TO THE BEACH, WITH ITS FOSSILS, SHELLS AND BUNCHES OF ARRANGED SEAWEED.

A FREQUENT CUSTOMER TO THE SHOP WAS LIEUTENANT-COLONEL THOMAS JAMES BIRCH.

BIRCH WAS A RETIRED OFFICER IN THE LIFE GUARDS WHO BOUGHT MANY FINE SPECIMENS FROM THE ANNINGS FOR HIS COLLECTION.

IN 1819 HE BECAME CONCERNED ABOUT THE FAMILY'S POVERTY. THEY HAD NOT FOUND A GOOD FOSSIL FOR A YEAR, AND, BECAUSE OF THIS, WERE ABOUT TO SELL THEIR FURNITURE TO PAY THE RENT.

BIRCH DECIDED HE WOULD AUCTION OFF THE FOSSILS HE'D BOUGHT FROM THE ANNINGS ON THEIR BEHALF.

IN A LETTER TO THE PALAEONTOLOGIST GIDEON MANTELL, HE WROTE...
I AM GOING TO SELL MY COLLECTION FOR THE BENEFIT OF THAT POOR WOMAN...

AND HER SON AND DAUGHTER AT LYME, WHO HAVE, IN TRUTH, FOUND ALMOST ALL THE FINE THINGS WHICH HAVE BEEN SUBMITTED TO SCIENTIFIC INVESTIGATION.

I MAY NEVER AGAIN POSSESS WHAT I AM ABOUT TO PART WITH, YET, IN DOING IT, I SHALL HAVE THE SATISFACTION OF KNOWING THE MONEY WILL BE WELL APPLIED.

THE SALE BROUGHT IN ABOUT 400 POUNDS, WHICH WAS ENOUGH TO PUT THE ANNINGS ON A GOOD FINANCIAL FOOTING, AND RAISE THEIR PROFILE WITHIN THE GEOLOGICAL COMMUNITY.

SO NAMED BECAUSE THEY WERE LIKE THE CONCRETIONS FOUND IN THE GALL BLADDER OF THE BEZOAR GOAT.

MARY ANNING HAD SEEN SUCH STONES INSIDE ICHTHYOSAURS, USUALLY IN THE ABDOMEN, NEAR THE RIBCAGE OR PELVIS.

SHE PUT FORWARD THE THEORY THAT BEZOAR STONES WERE THE FOSSILISED FAECES OF PREHISTORIC ANIMALS.

WILLIAM BUCKLAND AGREED. IN 1829 HE PUBLISHED THE FIRST PAPER ON WHAT WERE TO BE KNOWN AS COPROLITES.

MUCH COULD BE DISCOVERED ABOUT THE EATING HABITS OF THESE LONG-DEAD CREATURES BY CUTTING OPEN COPROLITES.
DISPERSED IRREGULARLY AND ABUNDANTLY THROUGHOUT THE FAECES...

ARE THE SCALES AND OCCASIONALLY THE TEETH AND BONES OF FISH.

MARY ANNING WAS MORE THAN JUST A FINDER OF FOSSILS, A SHOPKEEPER OR A GUIDE FOR GENTLEMAN GEOLOGISTS.

SHE WAS AS KNOWLEDGEABLE AS THE SCIENTISTS SHE DID BUSINESS WITH.

YET IT WAS ALWAYS WELL-EDUCATED MEN WHO PUBLISHED SCIENTIFIC DESCRIPTIONS (FORMAL DESCRIPTIONS OF A NEW SPECIES), OFTEN FAILING TO GIVE ANNING CREDIT FOR HER FINDS.

HER FRIEND ANNA PINNEY, WHO WAS A COLLECTOR HERSELF AND WHO OCCASIONALLY ACCOMPANIED MARY ALONG THE SHORELINE, WROTE...

SHE SAYS THE WORLD HAS USED HER ILL. THESE MEN OF LEARNING HAVE SUCKED HER BRAINS...

AND MADE A GREAT DEAL OF PUBLISHING WORKS OF WHICH SHE FURNISHED THE CONTENTS, WHILE SHE DERIVED NONE OF THE ADVANTAGES.

MARY DEVELOPED HER OWN SKILLS IN DISSECTION AND PRESERVATION. SHE WOULD DISMEMBER EXISTING CREATURES IN ORDER TO COMPARE THEM WITH THE FOSSILS SHE FOUND.

WHEN IN 1829 SHE DISCOVERED THE STRANGE FOSSIL FISH 'SQUALORAJA', EXPERTS THOUGHT IT WAS A RAY.

MARY DISSECTED A MODERN RAY AND FOUND THAT THE TWO ANIMALS WERE QUITE DIFFERENT. THIS WAS A SPECIES NEW TO SCIENCE.

SHE WOULD TAKE GREAT CARE AND TIME IN PRESERVING AND PRESENTING FOSSILS, SO THEY WOULD BE AS COMPLETE AS POSSIBLE FOR SCHOLARS TO STUDY.

A VISITOR TO LYME IN 1824, LADY HARRIET SILVESTER, HAD DESCRIBED MARY IN HER DIARY.

THE EXTRAORDINARY THING ABOUT THIS YOUNG WOMAN IS THAT SHE HAS MADE HERSELF SO THOROUGHLY ACQUAINTED WITH THE SCIENCE THAT THE MOMENT SHE FINDS ANY BONES SHE KNOWS TO WHAT TRIBE THEY BELONG.

SHE FIXES THE BONES ON A FRAME WITH CEMENT AND THEN MAKES DRAWINGS OF THEM AND HAS THEM ENGRAVED.

BY READING AND APPLICATION SHE HAS ARRIVED AT THE DEGREE OF KNOWLEDGE AS TO BE IN THE HABIT OF WRITING AND TALKING TO PROFESSORS AND OTHER CLEVER MEN ON THE SUBJECT.

EVERY MORNING, MARY ANNING WOULD GO WALKING UNDER THE CLIFFS AT LYME, ACCOMPANIED BY HER FAITHFUL DOG, TREY...

IN ORDER TO SEE WHAT FOSSILS HAD BEEN BROUGHT TO LIGHT BY THE FALLS OF ROCKS OR THE ACTION OF THE SEA.

ANNA PINNEY WROTE ABOUT HER FRIEND...
SHE SUPPORTED HER MOTHER AND BROTHER IN BITTER POVERTY, AND EVEN WHEN SHE WAS ILL, SHE WAS BROUGHT BACK FAINTING FROM THE BEACH.

FOSSIL-HUNTING COULD BE A DANGEROUS BUSINESS.

ONE DAY A ROCKFALL NARROWLY MISSED ANNING...

BUT KILLED POOR TREY.

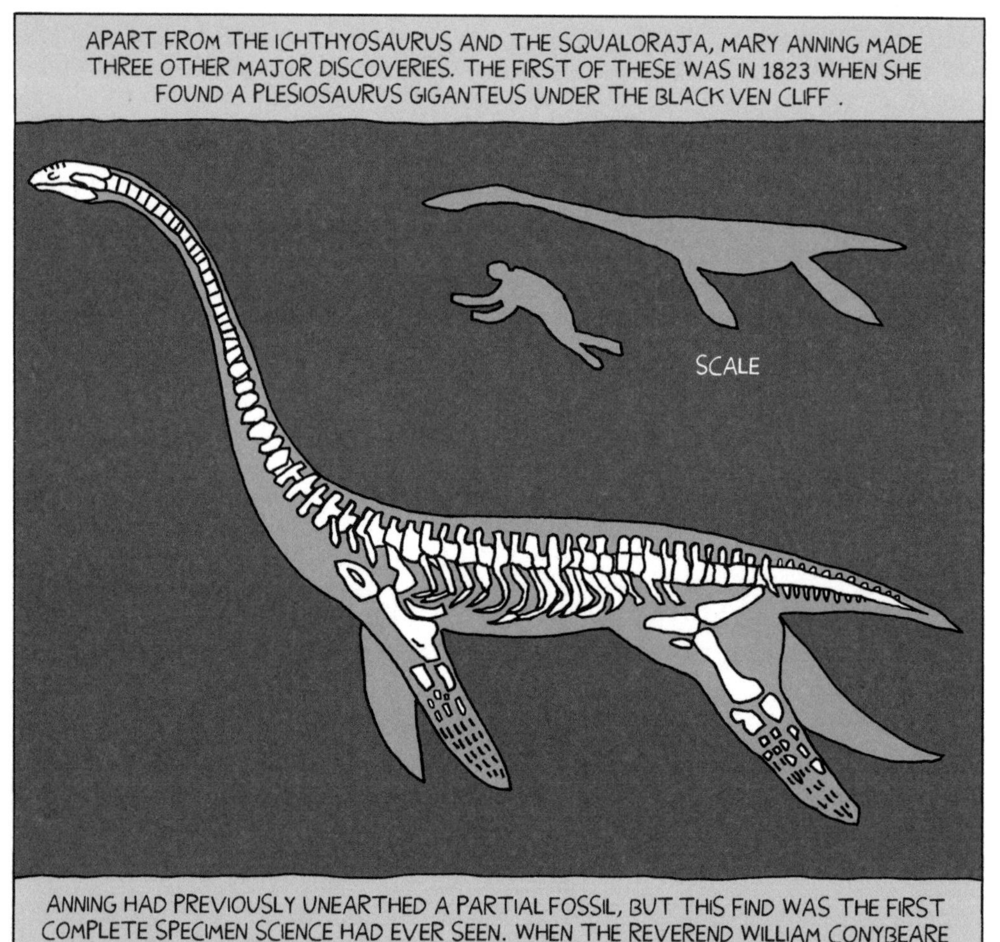
APART FROM THE ICHTHYOSAURUS AND THE SQUALORAJA, MARY ANNING MADE THREE OTHER MAJOR DISCOVERIES. THE FIRST OF THESE WAS IN 1823 WHEN SHE FOUND A PLESIOSAURUS GIGANTEUS UNDER THE BLACK VEN CLIFF.
SCALE
ANNING HAD PREVIOUSLY UNEARTHED A PARTIAL FOSSIL, BUT THIS FIND WAS THE FIRST COMPLETE SPECIMEN SCIENCE HAD EVER SEEN. WHEN THE REVEREND WILLIAM CONYBEARE GAVE A PRESENTATION ON THE ANIMAL AT THE GEOLOGICAL SOCIETY, HE FAILED TO MENTION MARY ANNING, AND WAS ULTIMATELY GIVEN CREDIT FOR THE FIND.

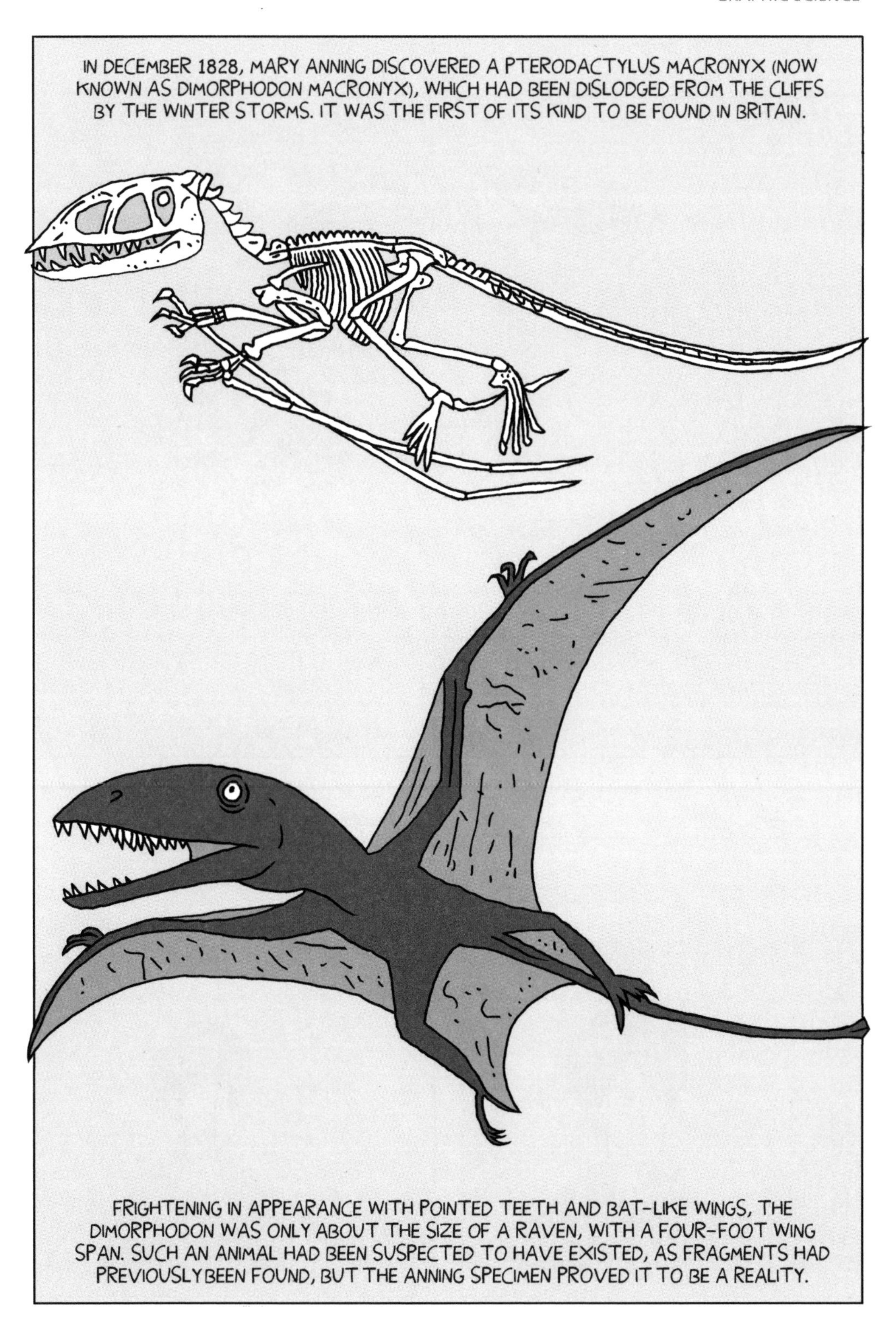
IN DECEMBER 1828, MARY ANNING DISCOVERED A PTERODACTYLUS MACRONYX (NOW KNOWN AS DIMORPHODON MACRONYX), WHICH HAD BEEN DISLODGED FROM THE CLIFFS BY THE WINTER STORMS. IT WAS THE FIRST OF ITS KIND TO BE FOUND IN BRITAIN.
FRIGHTENING IN APPEARANCE WITH POINTED TEETH AND BAT-LIKE WINGS, THE DIMORPHODON WAS ONLY ABOUT THE SIZE OF A RAVEN, WITH A FOUR-FOOT WING SPAN. SUCH AN ANIMAL HAD BEEN SUSPECTED TO HAVE EXISTED, AS FRAGMENTS HAD PREVIOUSLY BEEN FOUND, BUT THE ANNING SPECIMEN PROVED IT TO BE A REALITY.

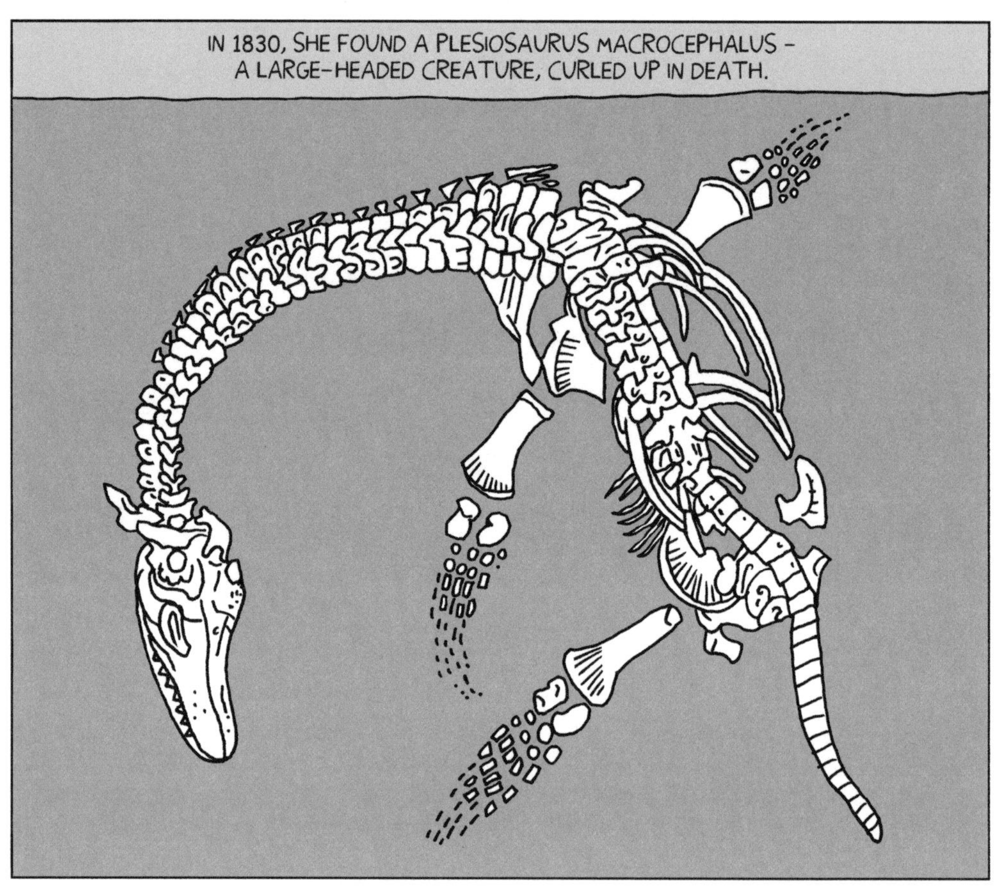
IN 1830, SHE FOUND A PLESIOSAURUS MACROCEPHALUS - A LARGE-HEADED CREATURE, CURLED UP IN DEATH.

SHE WROTE TO BUCKLAND...
THE HEAD IS TWICE AS LARGE IN PROPORTION AS THOSE I HAVE HITHERTO FOUND.

THE NECK HAS A MOST GRACEFUL CURVE, AND WHAT MAKES IT SO INTERESTING IS THAT RESTING ON THE BONES OF THE PELVIS IS ITS COPROLITE.

IN 1835 THE BRITISH ASSOCIATION FOR THE ADVANCEMENT OF SCIENCE GENERATED 200 POUNDS FOR HER SUPPORT THROUGH SUBSCRIPTION, AN AMOUNT TO WHICH THE BRITISH GOVERNMENT ADDED A FURTHER 300 POUNDS.

WHAT WAS MARY ANNING'S LEGACY? ALONG WITH WILLIAM BUCKLAND, WILLIAM CONYBEARE, GIDEON MANTELL AND OTHERS, SHE GAVE THE LATE GEORGIANS AND EARLY VICTORIANS A GLIMPSE INTO ANOTHER WORLD...
(THIS PANEL AFTER HENRY DE LA BECHE.)

ONE THAT WAS AT ODDS WITH THE BIBLICAL STORY OF CREATION. HOW DID SHE SQUARE HER RELIGIOUS BELIEFS WITH THE NEW SCIENCES OF PALAEONTOLOGY AND GEOLOGY?

HER CONCLUSIONS ON THIS SUBJECT ARE UNKNOWN, AS SHE MADE NO WRITTEN ACCOUNT OF HER THOUGHTS.

THAT SHE WAS ABLE TO CONTINUE HER WORK WITHOUT ANY APPARENT INTERNAL CONFLICT...

SUGGESTS THAT SHE FOUND A WAY OF INTEGRATING THE TWO WORLD VIEWS.

IN THIS, MARY ANNING WAS NO DIFFERENT FROM THE MANY GENTLEMAN CLERICS WHO SO ENTHUSIASTICALLY EXPLORED SCIENCE IN THE 19TH CENTURY.

FOR THEM, SCIENCE AND RELIGION WERE NOT IN OPPOSITION.

THE NATURAL PROCESSES THEY WITNESSED IN THEIR INVESTIGATIONS WERE SIMPLY THE WAY GOD WORKED IN THE PHYSICAL WORLD.

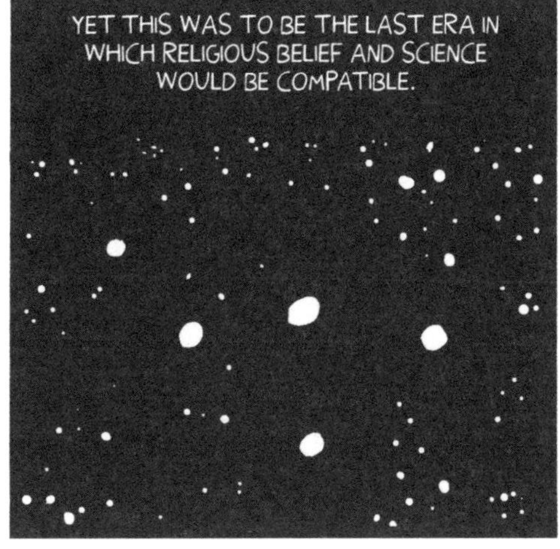
YET THIS WAS TO BE THE LAST ERA IN WHICH RELIGIOUS BELIEF AND SCIENCE WOULD BE COMPATIBLE.

THE PUBLISHING OF CHARLES DARWIN'S
*ON THE ORIGIN OF THE SPECIES* IN 1859...

OPENED A CHASM BETWEEN
SCIENCE AND RELIGION.

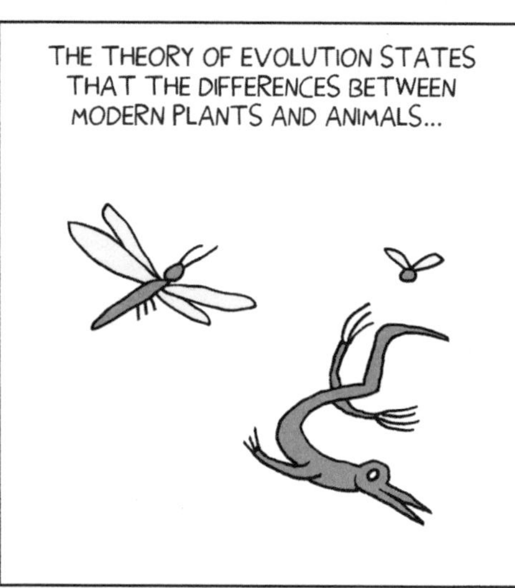
THE THEORY OF EVOLUTION STATES
THAT THE DIFFERENCES BETWEEN
MODERN PLANTS AND ANIMALS...

ARE DUE TO CHANGES THAT HAPPENED
BY NATURAL PROCESSES OVER A LONG
PERIOD OF TIME.

THAT SPAN OF TIME BEING BILLIONS OF
YEARS RATHER THAN THE FEW
THOUSAND YEARS FOR WHICH BIBLICAL
SCHOLARS BELIEVED THE EARTH TO
HAVE EXISTED.

FROM THE 19TH CENTURY ONWARDS,
SCIENCE BEGAN TO MOVE AWAY
FROM GOD.

MARY ANNING DIED OF BREAST CANCER AT THE AGE OF 47.

IN THE WORDS OF THE THEN PRESIDENT OF THE GEOLOGICAL SOCIETY, HENRY DE LA BECHE...
SHE BORE WITH FORTITUDE THE PROGRESS OF A CANCER ON HER BREAST UNTIL SHE FINALLY SUNK BENEATH ITS RAVAGES.

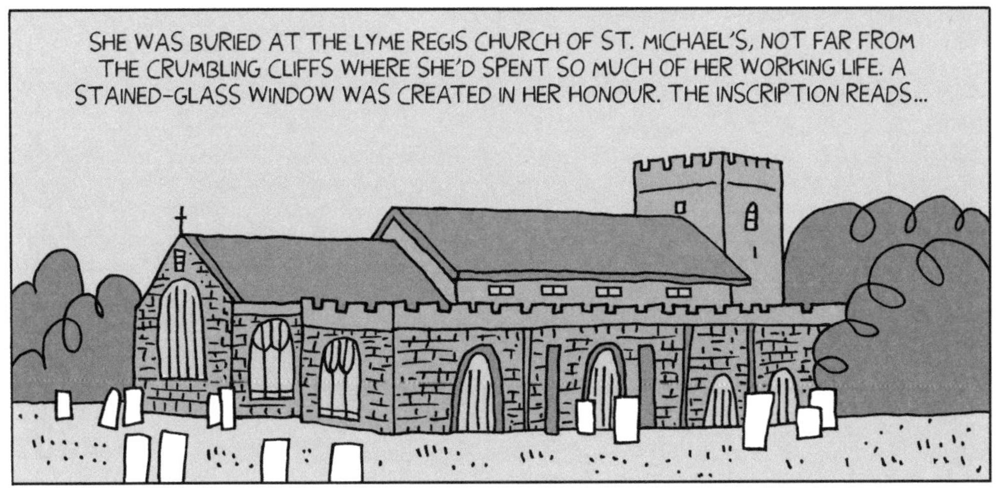
SHE WAS BURIED AT THE LYME REGIS CHURCH OF ST. MICHAEL'S, NOT FAR FROM THE CRUMBLING CLIFFS WHERE SHE'D SPENT SO MUCH OF HER WORKING LIFE. A STAINED-GLASS WINDOW WAS CREATED IN HER HONOUR. THE INSCRIPTION READS...

THIS WINDOW IS SACRED TO THE MEMORY OF MARY ANNING OF THIS PARISH, WHO DIED 9 MARCH AD 1847, AND IS ERECTED BY THE VICAR AND SOME MEMBERS OF THE GEOLOGICAL SOCIETY OF LONDON IN COMMEMORATION OF HER USEFULNESS IN FURTHERING THE SCIENCE OF GEOLOGY, AS ALSO OF HER BENEVOLENCE OF HEART AND INTEGRITY OF LIFE.

GEORGE
WASHINGTON
CARVER

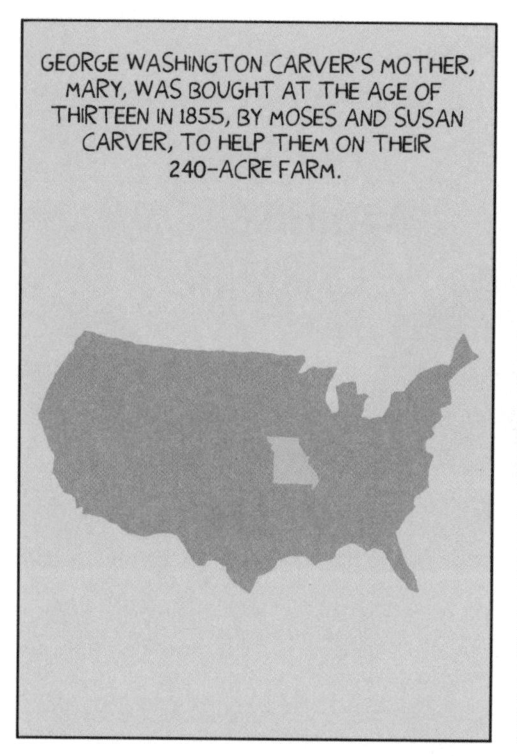
GEORGE WASHINGTON CARVER'S MOTHER, MARY, WAS BOUGHT AT THE AGE OF THIRTEEN IN 1855, BY MOSES AND SUSAN CARVER, TO HELP THEM ON THEIR 240-ACRE FARM.

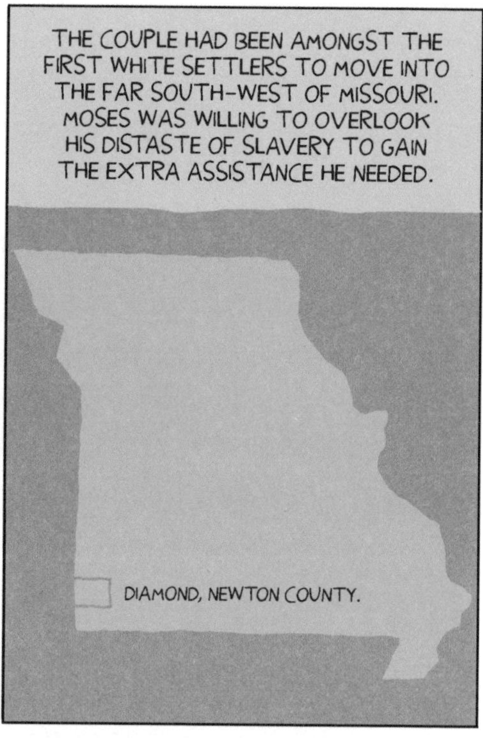
THE COUPLE HAD BEEN AMONGST THE FIRST WHITE SETTLERS TO MOVE INTO THE FAR SOUTH-WEST OF MISSOURI. MOSES WAS WILLING TO OVERLOOK HIS DISTASTE OF SLAVERY TO GAIN THE EXTRA ASSISTANCE HE NEEDED.
DIAMOND, NEWTON COUNTY.

MARY FIRST HAD A SON CALLED JAMES; THEN A SECOND SON, GEORGE. GEORGE'S FATHER, WHO DIED IN AN ACCIDENT BEFORE HIS CHILD WAS BORN, WAS THE PROPERTY OF A NEIGHBOURING FARMER.

THE EXACT DATE OF HIS BIRTH IS NOT KNOWN, BUT IT WAS POSSIBLY 1864 OR 1865, MAKING HIM ONE OF THE LAST AMERICANS TO BE BORN INTO SLAVERY.

THE CIVIL WAR WAS ENDING, PERHAPS HAD ENDED, WHEN RAIDERS ATTACKED THE CARVER FARM. THE OLDER BOY WAS SUCCESSFULLY HIDDEN, BUT GEORGE AND HIS MOTHER WERE TAKEN.

THE RAIDERS SOLD THE SLAVES IN ARKANSAS. MOSES CARVER HIRED A MAN TO FIND MARY AND THE INFANT...

BUT ONLY THE BABY WAS RETURNED. MARY COULD NOT BE FOUND.
THANK THE LORD!

THE RECOVERY OF GEORGE COST MOSES A HORSE WORTH 300 DOLLARS...
YOUR BROTHER IS HERE.

A CONSIDERABLE AMOUNT OF MONEY IN A TIME WHEN THE STATE OF MISSOURI HAD EITHER ABOLISHED SLAVERY OR WAS ABOUT TO.

THE ADULT GEORGE SPENT YEARS TRYING TO FIND HIS MOTHER. HIS MOST PRIZED POSSESSIONS WERE A BILL OF SALE THAT MADE HIS MOTHER THE PROPERTY OF THE CARVERS, AND A SPINNING WHEEL THAT HAD BELONGED TO HER.

DID HE AND HIS BROTHER, JAMES, HAVE A HAPPY CHILDHOOD AS SURROGATE SONS OF THE CARVERS? IT SEEMS THEY DID.

BOTH BOYS KEPT THEIR SURNAMES INTO ADULTHOOD. JAMES STAYED WITH THE CARVERS FOR TEN YEARS AFTER SLAVERY ENDED.

ACCORDING TO ONE BIOGRAPHER, GEORGE RETURNED TO NEWTON COUNTY NUMEROUS TIMES TO VISIT HIS FORMER OWNERS.

HE WOULD SOMETIMES BUY OVERALLS AND WORK SHIRTS TO BE SENT TO MOSES CARVER, AND OCCASIONALLY SEND POSTAL MONEY ORDERS FOR SMALL AMOUNTS OF MONEY, USUALLY FIVE OR TEN DOLLARS.

A SICKLY CHILD, GEORGE WAS UNABLE TO HELP MOSES IN THE FIELDS AS HIS OLDER BROTHER DID. INSTEAD HE HELPED SUSAN CARVER WITH HOUSEHOLD CHORES...

WHICH INCLUDED LAUNDRY, COOKING, CLEANING AND GARDENING.

PLANTS WERE OF PARTICULAR INTEREST TO HIM. HE WOULD SPEND HOURS IN THE WOODS ALONE LOOKING FOR FLORA.

OFTEN HE WOULD TRANSFER THESE FLOWERS INTO A SMALL HIDDEN GARDEN HE MANAGED OUT OF SIGHT OF THE CARVERS.

GEORGE FOUND HE HAD A TALENT FOR MAKING PLANTS THRIVE.

BUT HE KNEW THE NEIGHBOURHOOD WOULD CONSIDER IT FOOLISH TO WASTE TIME ON FLOWERS.

GEORGE WASHINGTON CARVER HAD A STRONG BELIEF IN GOD, WHO HE WAS CONVINCED SPOKE TO HIM IN DREAMS.

THROUGHOUT HIS LIFE HE WOULD TELL THE STORY OF WHEN, AS A BOY, HE HAD LONGED FOR A POCKET KNIFE OF HIS OWN.

ONE NIGHT, HE DREAMED THAT HE SAW THIS DESIRED ITEM IN A FIELD NEARBY, STICKING OUT OF A PARTIALLY EATEN WATERMELON.

THE NEXT MORNING HE WALKED TO THAT SPOT ON THE CARVER FARM...

AND FOUND...

THE VERY THING HE'D DREAMED OF.

GEORGE'S LIFE IN THE POST-CIVIL-WAR ERA STANDS IN SHARP CONTRAST TO THE EXPERIENCES OF MANY AFRICAN AMERICANS AT THIS TIME.

THE NORTHERN GOVERNMENT FAILED TO PUT SYSTEMS IN PLACE TO EASE THE SUFFERING OF THE MANY NEWLY FREED PEOPLE.

FORMER SLAVES WERE OFTEN NEGLECTED BY UNION SOLDIERS OR FACED RAMPANT DISEASE, WHICH INCLUDED OUTBREAKS OF SMALLPOX AND CHOLERA. MANY SIMPLY STARVED TO DEATH.

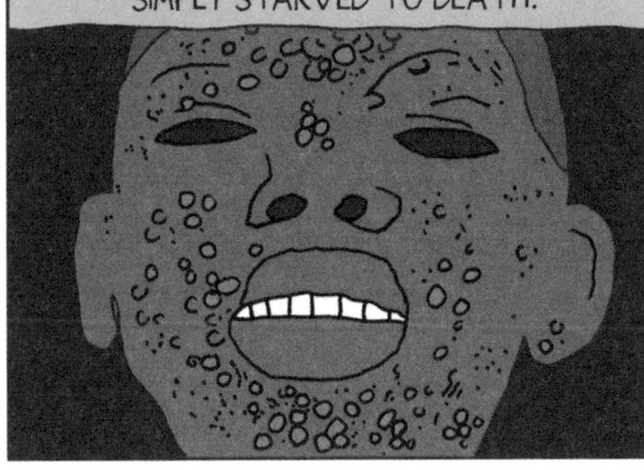

THE SICKNESS AND DEATH RATES WERE SO HIGH THAT SOME COMMENTATORS WONDERED IF BLACK AMERICANS WOULD DIE OUT.

THE REVEREND J. STURTEVANT WROTE IN 1863...

LIKE HIS BRETHREN THE INDIAN IN THE FOREST, HE MUST MELT AWAY AND DISAPPEAR FOREVER FROM THE REST OF US.

SUCH THEORIES GENERATED A CONVENIENT RATIONALE FOR OPPRESSIVE NEW RACIAL POLICIES.

FINDING THAT THE SCHOOL WAS CLOSED, HE SLEPT IN A BARN OVERNIGHT.

CARVER DID NOT STAY LONG IN NEOSHO – PERHAPS LESS THAN A YEAR. WHILE STILL A TEENAGER, HE JOINED THE FIRST GREAT MIGRATION OF AFRICAN AMERICANS OUT OF THE SOUTH.

HE HITCHED A RIDE WTH A BLACK FAMILY WHO WERE EN ROUTE TO KANSAS. THEY WERE FLEEING THE SOUTHERN STATES TO AVOID THE CONSEQUENCES OF THE 'COMPROMISE OF 1877'.

THIS WAS A POLITICAL DEAL THAT ALLOWED THE REPUBLICAN RUTHERFORD B. HAYES INTO THE WHITE HOUSE IN RETURN FOR THE FEDERAL GOVERNMENT PULLING THE REMAINING NORTHERN TROOPS FROM THE SOUTH.

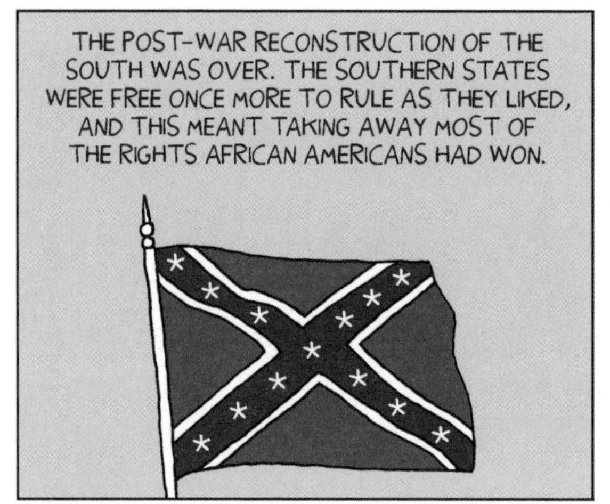

BY THE BEGINNING OF THE 20TH CENTURY, MOST BLACK PEOPLE WERE DISENFRANCHISED BY STATE LEGISLATORS IN EVERY SOUTHERN STATE, DESPITE BEING THE MAJORITY IN SOME.

THEY MAY HAVE ASSUMED THAT WHITE PEOPLE OUTSIDE THE SOUTH WERE LESS PRONE TO BE PREJUDICED AGAINST BLACKS.

CARVER FOUND HIMSELF IN FORT SCOTT, KANSAS...

A TOWN THAT SAW ITS AFRICAN-AMERICAN POPULATION GROW FROM 682 IN 1870 TO 1,082 IN 1880 – A SITUATION THAT NO DOUBT CONTRIBUTED TO RACIAL TENSIONS.

WHILE IN FORT SCOTT, CARVER WITNESSED THE LYNCHING OF A BLACK MAN ACCUSED OF RAPING A WHITE WOMAN.

THIS MAN WAS TAKEN BY FORCE FROM THE COUNTY JAIL BY A GROUP OF MASKED MEN.
JAIL

THEN, IN FRONT OF THOUSANDS OF JEERING TOWNSPEOPLE, HE WAS FIRST DRAGGED BY THE NECK DOWN THE STREET...

THEN, HALF-DEAD ALREADY, HE WAS HANGED...

AFTER WHICH HIS BODY WAS BURNED IN A PUBLIC ACT OF DESECRATION.

CARVER LEFT TOWN IMMEDIATELY.

SOME YEARS OF WANDERING THEN BEGAN...
DRY GOODS

BOARDING WITH BLACK FAMILIES, WORKING ODD JOBS, GETTING SCHOOLING WHEN HE COULD...

IN OLATHE, AND PAOLA, THEN MINNEAPOLIS, KANSAS.

AROUND THIS TIME, CARVER ADDED THE MIDDLE INITIAL 'W' TO HIS NAME IN ORDER TO ENSURE HIS MAIL DIDN'T GO TO ANOTHER GEORGE CARVER WHO LIVED IN THE VICINITY.

MAIL FOR YOU.

HOWEVER, ON SEEING HIM...
WE DON'T ACCEPT COLOUREDS HERE.

CARVER'S NEXT STOP WAS EDEN TOWNSHIP, IN NESS COUNTY, KANSAS, WHERE HE STAKED A CLAIM UNDER THE HOMESTEAD ACT.

HERE HE BUILT HIS OWN SOD HOUSE, WHICH HE MOVED INTO ON 20TH APRIL 1888.

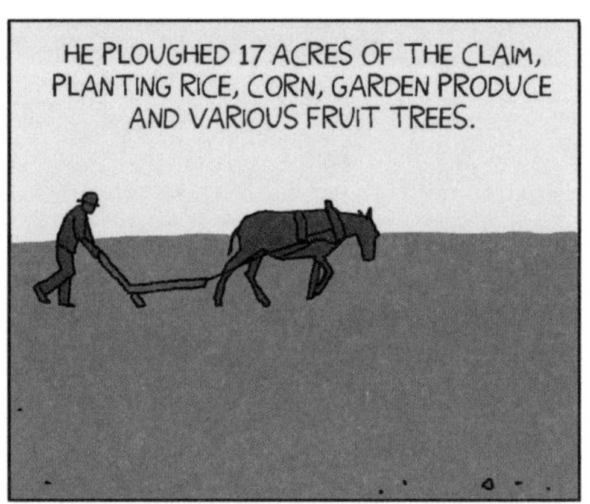
HE PLOUGHED 17 ACRES OF THE CLAIM, PLANTING RICE, CORN, GARDEN PRODUCE AND VARIOUS FRUIT TREES.

DESPITE ALL THIS BACK-BREAKING WORK, THE CLAIM DID NOT THRIVE. CARVER WAS UNABLE TO FIND A WATER SOURCE ON THE LAND...

AND HIS ATTEMPT AT FARMING COINCIDED WITH ONE OF THE WORST DROUGHTS AND THEN HARSHEST WINTERS IN KANSAS HISTORY.

CARVER MOVED TO INDIANOLA, IOWA, WHERE HE ENROLLED AT SIMPSON COLLEGE, STUDYING ART AND MUSIC.

RELYING ONCE MORE ON SKILLS HE'D LEARNED AS A CHILD, HE OPENED A LAUNDRY TO EARN MONEY.

AS THE ONLY AFRICAN AMERICAN AT THE COLLEGE, CARVER FOUND HIMSELF THE FOCUS OF MUCH INTEREST FROM THE OTHER STUDENTS. WHO WAS THIS QUIETLY SPOKEN AND OBVIOUSLY INTELLIGENT BLACK MAN?

WHEN IT WAS NOTICED THAT CARVER HAD NO FURNITURE, A COLLECTION WAS TAKEN UP TO PROVIDE A TABLE, CHAIRS AND A BED.

THIS FURNITURE WAS THEN SNEAKED INTO CARVER'S QUARTERS WHEN HE WAS OUT ONE DAY.
!

STUDENTS WOULD ALSO OCCASIONALLY LEAVE HIM MONEY ANONYMOUSLY.
I OFTEN CAME HOME AND FOUND 25 CENTS OR 50 CENTS UNDER MY DOOR.

CARVER'S ART TEACHER, ETTA BUD, RECOGNISED HER STUDENT'S TALENT IN PAINTING, BUT WORRIED HE WOULDN'T BE ABLE TO MAKE A LIVING AS A BLACK ARTIST..

SHE SUGGESTED HE SHOULD CONSIDER STUDYING BOTANY AT THE IOWA STATE AGRICULTURE COLLEGE. THIS HE DID, ONCE AGAIN FINDING HIMSELF THE LONE AFRICAN AMERICAN ON THE CAMPUS.

CARVER QUICKLY MADE AN IMPRESSION ON THE FACULTY AT IOWA STATE COLLEGE. SUCH WAS HIS SKILL AT RAISING, CROSS-FERTILISING AND GRAFTING PLANTS THAT HE WAS ENCOURAGED TO STAY ON AFTER HIS SENIOR YEAR AS A GRADUATE STUDENT. HE WORKED ALONGSIDE NOTED FUNGUS EXPERT L.H. PALMER, HONING HIS TALENTS AT TREATING PLANT DISEASES.

HE OBTAINED HIS MASTER OF AGRICULTURE DEGREE IN 1896. HE IMMEDIATELY RECEIVED A NUMBER OF JOB OFFERS, INCLUDING ONE FROM TUSKEGEE INSTITUTE IN ALABAMA. THIS COLLEGE OF EDUCATION, RUN BY EDUCATOR, AUTHOR, ORATOR AND NATIONAL LEADER AMONG AFRICAN AMERICANS BOOKER T. WASHINGTON, WAS A TEACHER-TRAINING UNIVERSITY FOR BLACKS.

A DECADE OLDER THAN CARVER, BOOKER T. WASHINGTON HAD A CLEAR MEMORY OF SLAVERY.

HE'D ESTABLISHED TUSKEGEE AS AN EDUCATIONAL ESTABLISHMENT FOR AFRICAN AMERICANS TO LEARN INDUSTRIAL AND FARMING SKILLS, AND THE CULTIVATION OF THE VIRTUES OF ENTERPRISE AND THRIFT.

IN 1895 WASHINGTON MADE A SPEECH SETTING OUT HIS PHILOSOPHY OF ACCOMMODATION...

STATING THAT AFRICAN AMERICANS SHOULD GAIN ECONOMIC INDEPENDENCE BEFORE SEEKING SOCIAL AND POLITICAL EQUALITY.

THEY SHOULD LEARN TO SUPPORT THEMSELVES AND THEIR FAMILIES AND ACCEPT THE SOUTH'S RACIAL SYSTEM UNTIL THEY HAD PROVEN THEMSELVES.
REST ROOMS
WHITE
COLORED

CARVER APPARENTLY APPROVED OF THIS STRATEGY. HE WROTE TO WASHINGTON, SAYING...
I READ YOUR STIRRING ADDRESS DELIVERED AT CHICAGO AND I SAID AMEN TO ALL YOU SAID. YOU HAVE THE RIGHT SOLUTION TO THE RACE PROBLEM.

NOT EVERYONE AGREED. WASHINGTON'S VIEWS CAUSED CONTROVERSY AMONG AFRICAN AMERICANS. W.E.B. DU BOIS, THE OTHER GREAT BLACK LEADER OF THAT ERA, STRONGLY DISAGREED, BELIEVING THAT THIS STRATEGY WOULD ONLY PERPETUATE WHITE OPPRESSION. HE WANTED FULL CIVIL RIGHTS AND INCREASED POLITICAL REPRESENTATION, NOT HALF-MEASURES DESIGNED TO APPEASE RACISTS.

GEORGE WASHINGTON CARVER ARRIVED AT TUSKEGEE CAMPUS IN 1896. HE WAS THE ONLY AFRICAN AMERICAN IN THE COUNTRY WITH A MASTERS DEGREE IN AGRICULTURE...

AND HE DIDN'T LET OTHERS FORGET IT.

IN THE EARLY DAYS OF THE INSTITUTE, UNMARRIED FACULTY MEMBERS NORMALLY HAD TO SHARE ROOMS.

NOT ONLY DID CARVER NOT HAVE TO SHARE, BUT HE INSISTED ON HAVING AN EXTRA ROOM JUST FOR HIS PLANT COLLECTION.

THIS, AND THE FACT THAT HE WAS PAID AN ABOVE-AVERAGE SALARY, CAUSED SOME RESENTMENT AMONG THE OTHER FACULTY MEMBERS.

CARVER'S DUTIES AT TUSKEGEE WERE EXTENSIVE. APART FROM HIS TEACHING COMMITMENTS, HE WAS ALSO IN CHARGE OF THE AGRICULTURAL EXPERIMENT STATION...

AND HE HAD TO MANAGE THE PRODUCTION AND SALE OF FARM PRODUCTS TO GENERATE INCOME FOR THE INSTITUTE.

ALTHOUGH A SKILLED AND POPULAR TEACHER, CARVER PROVED TO BE A POOR ADMINISTRATOR.

THIS WEAKNESS WAS EXACERBATED BY BOOKER T. WASHINGTON'S TENDENCY TO MICRO-MANAGE ALL ASPECTS OF THE INSTITUTION.

ONE TIME, CARVER WAS REPRIMANDED BY WASHINGTON FOR BEING LATE FOR A STAFF MEETING.

CARVER COUNTERED THAT HE WAS LATE ONLY BECAUSE...
I WAS DETAINED BY YOU IN YOUR OFFICE.

IN 1904 AN INSTITUTE COMMITTEE CLAIMED THAT CARVER'S REPORTS OF YIELDS FROM THE POULTRY YARD HAD BEEN EXAGGERATED.

NOW TO BE BRANDED AS A LIAR AND PARTY TO SUCH HELLISH DECEPTION IS MORE THAN I CAN BEAR.

IF YOUR COMMITTEE FEEL THAT I HAVE WILFULLY LIED OR PARTY TO SUCH LIES AS WERE TOLD, MY RESIGNATION IS AT YOUR DISPOSAL.

FRICTION BETWEEN THE TWO MEN WOULD CONTINUE FOR TWENTY YEARS, WITH CARVER FREQUENTLY THREATENING TO RESIGN, ALTHOUGH HE NEVER DID.

AREAS OF CONFLICT INCLUDED CARVER'S DESIRE TO DO LESS TEACHING SO HE COULD CONCENTRATE ON RESEARCH...

AND HIS WISH FOR A FULLY EQUIPPED LABORATORY AND EXTRA RESOURCES, WHILE WASHINGTON INSISTED HE DO MORE WITH LESS.

THEY HAD A DIFFICULT RELATIONSHIP, BUT THE TWO MEN DID APPEAR TO HAVE GREAT RESPECT FOR EACH OTHER.

BOOKER T. WASHINGTON DIED IN 1915, AFTER WHICH FEWER DEMANDS WERE MADE ON CARVER FOR ADMINISTRATIVE TASKS.
WASHINGTON


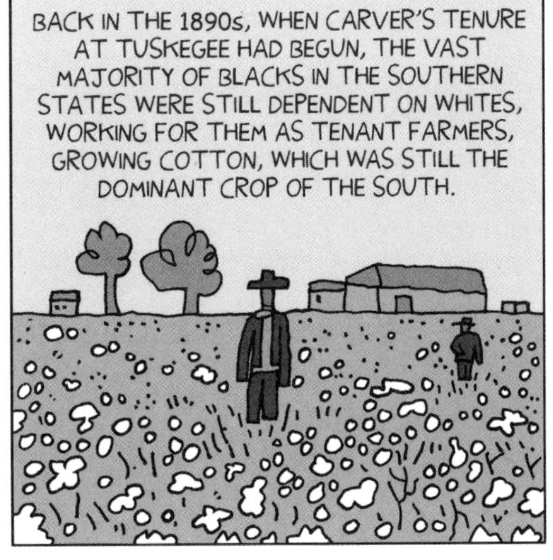

BACK IN THE 1890s, WHEN CARVER'S TENURE AT TUSKEGEE HAD BEGUN, THE VAST MAJORITY OF BLACKS IN THE SOUTHERN STATES WERE STILL DEPENDENT ON WHITES, WORKING FOR THEM AS TENANT FARMERS, GROWING COTTON, WHICH WAS STILL THE DOMINANT CROP OF THE SOUTH.

WHEN CARVER FIRST ARRIVED IN ALABAMA, HE WAS SHOCKED BY THE CONTRAST WITH THE GOLDEN WHEAT FIELDS AND TALL GREEN CORN HE'D LEFT IN IOWA.

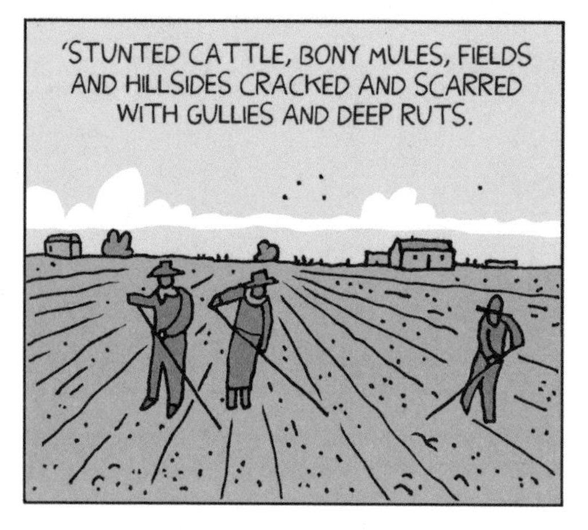

'STUNTED CATTLE, BONY MULES, FIELDS AND HILLSIDES CRACKED AND SCARRED WITH GULLIES AND DEEP RUTS.

'NOT MUCH EVIDENCE OF SCIENTIFIC FARMING ANYWHERE. EVERYTHING LOOKED HUNGRY.

'THE LAND, THE COTTON, THE CATTLE...

'AND THE PEOPLE.'

THE SINGLE GREATEST AGRICULTURAL PROBLEM WAS EROSION, AND COTTON WAS THE BIGGEST CONTRIBUTOR TO IT.

AT TUSKEGEE, CARVER BEGAN TO DEVELOP TECHNIQUES FOR IMPROVING SOILS DEPLETED BY REPEATED PLANTING OF COTTON.

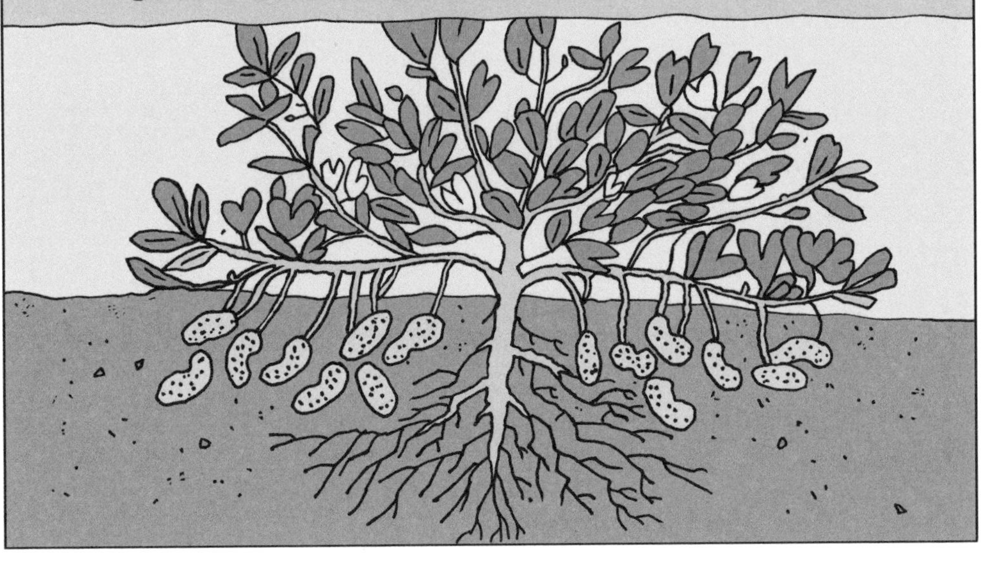
ALONG WITH OTHER AGRICULTURAL EXPERTS, CARVER ENCOURAGED FARMERS TO RESTORE NITROGEN TO THEIR SOIL BY PRACTISING CROP ROTATION, ALTERNATING COTTON CROPS WITH PLANTINGS OF SWEET POTATOES, SOYBEANS OR PEANUTS. BY USING THESE SOIL-ENRICHING PLANTS AND GOOD CULTIVATION METHODS, CARVER WAS ABLE TO INCREASE SOIL PRODUCTIVITY DRAMATICALLY.

CARVER HAD TO ACCOMPLISH THIS WITHOUT THE USE OF COMMERCIAL FERTILISERS, AS THIS WAS AN EXPENSE BEYOND THE REACH OF MOST POOR SOUTHERN FARMERS. HE WAS AWARE THAT EVERY OPERATION HE PERFORMED HAD TO BE WITHIN THE BUDGET OF 'POOR TENANT FARMERS WITH ONE-HORSE EQUIPMENT'.

AS THE DIRECTOR OF THE AGRICULTURAL EXPERIMENT STATION, CARVER PUBLISHED REGULAR BULLETINS THROUGH WHICH HE REACHED A WIDE AUDIENCE.

HIS READERSHIP WAS POOR FARMERS WHO WERE ONLY MARGINALLY EDUCATED, SO THE BULLETINS WERE WRITTEN IN A STRAIGHTFORWARD AND SIMPLE MANNER.

TITLES INCLUDED 'EXPERIMENTS WITH SWEETCORN', 'HOW TO BUILD UP WORN OUT SOILS' AND 'THREE DELICIOUS MEALS EVERY DAY FOR THE FARMER'.

CARVER ALSO OUTFITTED A TRAVELLING AGRICULTURE SCHOOL NAMED THE 'JESSOP WAGON', AFTER THE NEW YORK BANKER MORRIS K. JESSOP, WHO SUPPLIED THE FUNDING.

CARVER DIDN'T TRAVEL WITH THE WAGON HIMSELF, BUT HE DID DRAFT PLANS FOR IT, SELECTED THE EQUIPMENT, AND SUGGESTED LECTURES ON FARMING. THE MOBILE SCHOOL OPENED FOR BUSINESS ON 24TH MAY 1906, AND WITHIN A FEW YEARS WAS MADE PART OF THE U.S. DEPARTMENT OF AGRICULTURE'S OUTREACH PROGRAMME.

IT WAS THE HUMBLE PEANUT THAT BROUGHT CARVER NATIONAL FAME. IN 1921, THE WAYS AND MEANS COMMITTEE OF THE U.S. HOUSE OF REPRESENTATIVES HELD HEARINGS ON A PROPOSAL THAT WOULD PROTECT AMERICAN PEANUT-GROWERS AGAINST FOREIGN COMPETITION BY IMPOSING A TARIFF ON IMPORTED PEANUTS. CARVER WAS ASKED TO ADDRESS THE COMMITTEE AS AN EXPERT WITNESS.
THIS WAS UNUSUAL, AS THERE WERE NO AFRICAN AMERICANS IN CONGRESS AT THAT TIME.
BLACKS HELD ONLY MENIAL JOBS. RESTROOMS AND EATING FACILITIES IN THE BUILDING WERE RACIALLY SEGREGATED.

CARVER WAS ALLOTTED TEN MINUTES, BUT ENDED UP SPEAKING FOR FORTY-FIVE, AT WHICH POINT HE RECEIVED A STANDING OVATION.
KLAP! KLAP! KLAP!

HE HAD MADE THE CASE FOR THE PEANUT TARIFF AND ACHIEVED NATIONAL RECOGNITION.

CONTRARY TO POPULAR BELIEF, CARVER DID NOT INVENT PEANUT BUTTER. HE DID DEVELOP OVER 300 USES FOR PEANUTS, BUT FEW WERE FINANCIALLY SUCCESSFUL.

HOWEVER, HIS PROMOTION OF PEANUTS DID INCREASE THE POPULARITY OF THE FOODSTUFF, MAKING IT A STAPLE OF THE AMERICAN DIET.

CARVER WAS A SKILLED MASSEUR. IN THE 1930s HE BEGAN TO USE PEANUT OIL IN MASSAGE TO ALLEVIATE THE SYMPTOMS OF POLIO IN PATIENTS.

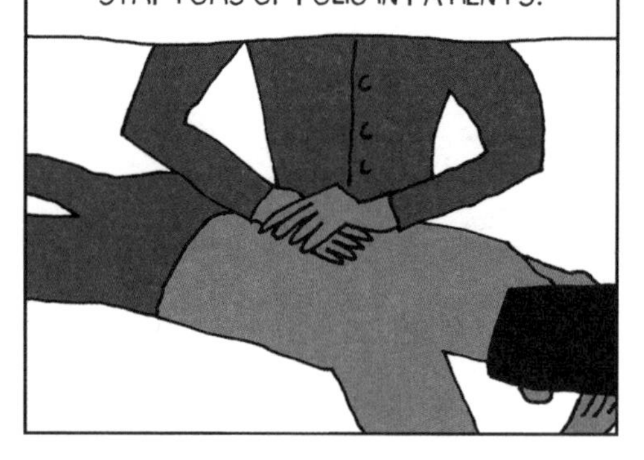

THE DISEASE, POLIO, FREQUENTLY CAUSED DAMAGE TO THE NERVE CELLS IN THE SPINAL CORD, CAUSING PARALYSIS AND GRADUAL MUSCLE-WASTING.

CARVER SAW THERE WAS IMPROVEMENT IN THE MUSCLES OF POLIO VICTIMS. HE BECAME CONVINCED THAT THE NUTRITIVE QUALITIES OF PEANUT OIL WERE BEING ABSORBED THROUGH THE SKIN INTO THE BLOODSTREAM.

HOWEVER, THERE IS NO SCIENTIFIC EVIDENCE THAT PEANUT OIL MASSAGE WORKS. IT'S MORE LIKELY THAT PATIENTS BENEFITED FROM THE MASSAGE ITSELF, RATHER THAN FROM ANY QUALITIES IN THE OIL.

CARVER NEVER MARRIED. AT THE AGE OF FORTY HE BEGAN A COURTSHIP WITH SARAH C. HUNT, AN ELEMENTARY SCHOOL TEACHER, BUT THE RELATIONSHIP DIDN'T LAST.

CARVER WAS A MAN WHO DEVELOPED STRONG BONDS WITH MANY OF HIS STUDENTS AND OTHER MALE ADMIRERS OF HIS WORK.

THERE IS NO EVIDENCE THAT CARVER EVER HAD A PHYSICAL RELATIONSHIP WITH ANY OF THEM, BUT THE POWERFUL LETTERS HE OCCASIONALLY SENT TO 'HIS BOYS' GIVES US SOME INSIGHT INTO HIS EMOTIONAL STATE.

THESE LETTERS CONTAIN A CURIOUS MIXTURE OF LONGING, LONELINESS AND RELIGIOUS FERVOUR.

IT IS MORE LIKELY THAT HE SAW 'HIS BOYS' AS SURROGATE CHILDREN FROM WHOM HE DERIVED A GREAT DEAL OF EMOTIONAL SATISFACTION...

AND THROUGH WHOM HE HOPED HIS WORK AND MESSAGE COULD BE CARRIED ON AFTER HIS DEATH.

IN A TIME WHEN SCIENCE WAS MOVING AWAY FROM RELIGION, CARVER'S DEVOUT CHRISTIANITY WAS UNUSUAL.

HIS SOMETIMES LESS-THAN-SCIENTIFIC APPROACH DREW CRITICISM FROM OTHER ACADEMICS.

BUT HE REMAINED RESOLUTE IN HIS FAITH THROUGHOUT HIS LIFE.

READING ABOUT NATURE IS FINE, BUT IF A PERSON WALKS IN THE WOODS AND LISTENS CAREFULLY...

HE CAN LEARN MORE THAN WHAT IS IN BOOKS...

FOR THEY SPEAK WITH THE VOICE OF GOD.

IN 1938, THE THEN PRESIDENT OF THE TUSKEGEE INSTITUTE AUTHORISED THE CREATION OF THE GEORGE WASHINGTON CARVER MUSEUM...

TO BE HOUSED IN WHAT WAS PREVIOUSLY THE SCHOOL LAUNDRY.
GEORGE WASHINGTON CARVER MUSEUM

THE MUSEUM DISPLAYED CARVER'S EXTENSIVE COLLECTION OF NATIVE PLANTS, MINERALS, BIRDS AND VEGETABLES...

AND ALSO HIS PAINTINGS, DRAWINGS AND TEXTILE ART.

THE MUSEUM WAS FORMALLY OPENED IN 1941 BY MR AND MRS HENRY FORD.

INCREASINGLY FRAIL IN HIS OLD AGE, CARVER FELL ONE DAY WHILE OPENING THE MUSEUM DOOR.

HE NEVER RECOVERED FROM THIS FALL AND DIED ON 5TH JANUARY 1943, AT TUSKEGEE INSTITUTE, WHERE HE HAD LIVED AND TAUGHT FOR ALMOST HALF A CENTURY.

GEORGE WASHINGTON CARVER HAS OFTEN BEEN HAILED AS THE 'PEANUT MAN', AND HAS COME TO SYMBOLISE THE INTELLECTUAL ACHIEVEMENT OF AFRICAN AMERICANS IN THE FACE OF PREJUDICE.

REAL BENEFITS CAME FROM HIS ENCOURAGEMENT OF IMPROVED FARM METHODS, SUCH AS CROP ROTATION AND SOIL CONSERVATION. THIS GAVE MILLIONS OF PEOPLE A CHANCE TO LIFT THEMSELVES OUT OF POVERTY.

NIKOLA
TESLA

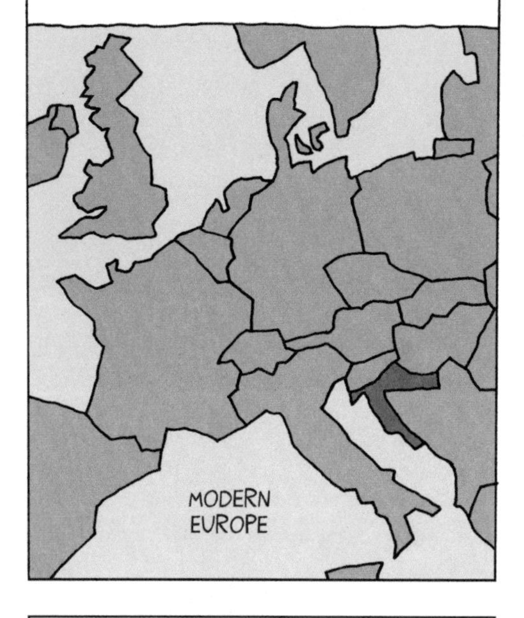
NIKOLA TESLA WAS BORN ON 10TH JULY 1858 IN SMILJAN, A SMALL VILLAGE IN THE MOUNTAINOUS REGION OF WHAT IS NOW CROATIA, BUT WAS THEN PART OF THE AUSTRO-HUNGARIAN EMPIRE.
MODERN EUROPE

HIS FAMILY WERE SERBIANS, A SOUTH SLAVIC ETHNIC GROUP NATIVE TO THE BALKANS.
SERBIA
SLOVENIA
CROATIA
BOSNIA AND HERZEGOVINA
SMILJAN
MODERN EUROPE

TESLA'S FATHER, MILUTIN, WAS THE VILLAGE ORTHODOX PRIEST.

HIS WIFE, DJOUKA, BORE FIVE CHILDREN, OF WHOM NIKOLA WAS THE FOURTH. THERE WERE TWO BOYS - DANE AND NIKOLA - AND THREE GIRLS - ANGELINA, MILKA AND MARICA.

THE CHILDREN LED AN IDYLLIC LIFE AROUND THE VILLAGE: SWIMMING, CATCHING FROGS IN THE SUMMER, AND DAMMING STREAMS IN THE WINTER.

THE BOYS' MOST INGENIOUS INVENTION WAS A PROPELLER...

DRIVEN BY SIXTEEN MAYFLY BUGS, GLUED OR SEWN ON TO THE WOODEN BLADES.

THESE CREATURES WERE REMARKABLY EFFICIENT, FOR ONCE THEY STARTED THEY HAD NO SENSE TO STOP AND CONTINUED WHIRLING FOR HOURS AND HOURS.

TRAGEDY STRUCK THE TESLA FAMILY WHEN THE ELDER BOY, DANE...

WAS THROWN BY A HORSE AND KILLED.

NIKOLA BEGAN HIS FORMAL EDUCATION HERE, WORKING HIS WAY THROUGH PRIMARY SCHOOL, AND THEN REAL GYMNASIUM (HIGH SCHOOL).

IN MY BOYHOOD I SUFFERED FROM A PECULIAR AFFLICTION DUE TO THE APPEARANCE OF IMAGES...

OFTEN ACCOMPANIED BY FLASHES OF LIGHT WHICH MARRED THE SIGHT OF REAL OBJECTS.

WHEN A WORD WAS SPOKEN TO ME, THE IMAGE OF THE OBJECT IT DESIGNATED WOULD PRESENT ITSELF VIVIDLY TO MY VISION.

I WONDER WHERE THE KETTLE IS?

THIS STRANGE TRICK OF THE MIND WORKED TO TESLA'S ADVANTAGE WHEN HIS THOUGHTS TURNED TO INVENTION.

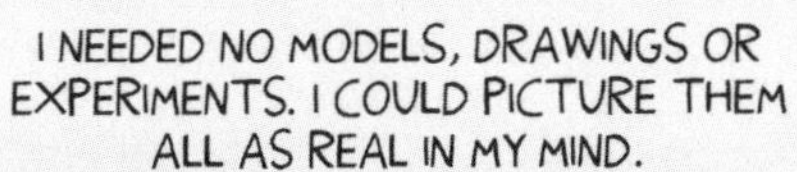

DURING THIS PERIOD I CONTRACTED MANY STRANGE LIKES, DISLIKES AND HABITS. I HAD A VIOLENT AVERSION TO THE EARRINGS OF WOMEN...

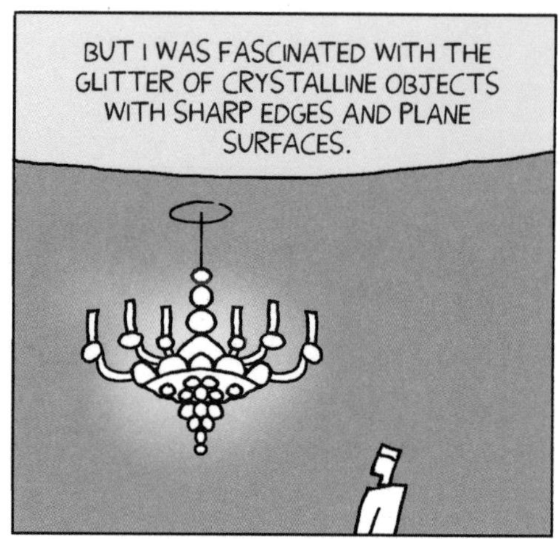
BUT I WAS FASCINATED WITH THE GLITTER OF CRYSTALLINE OBJECTS WITH SHARP EDGES AND PLANE SURFACES.

I WOULD NOT TOUCH THE HAIR OF OTHER PEOPLE, EXCEPT PERHAPS AT THE POINT OF A REVOLVER.

I WOULD GET A FEVER BY LOOKING AT A PEACH.
UGH!

I COUNTED THE STEPS OF MY WALKS...

AND CALCULATED THE CUBICAL VOLUME OF SOUP PLATES, COFFEE CUPS AND PIECES OF FOOD.

OTHERWISE MY MEAL WAS UNENJOYABLE.
YOU'RE NOT EATING, NIKO.

ALL REPEATED ACTS OR OPERATIONS I PERFORMED HAD TO BE DIVISIBLE BY THREE, AND, IF I MISSED, I FELT IMPELLED TO DO THEM ALL OVER AGAIN, EVEN IF IT TOOK HOURS.

I WAS INTERESTED IN ELECTRICITY FROM THE VERY BEGINNING OF MY EDUCATIONAL CAREER.

I READ ALL THAT I COULD ON THE SUBJECT, AND EXPERIMENTED WITH BATTERIES AND INDUCTION COILS.

IN 1870, AT THE AGE OF FOURTEEN, TESLA MOVED FROM GOSPIC TO KARLOVAC TO ATTEND HIGHER GYMNASIUM.

HERE HE SAW HIS FIRST STEAM TRAIN.

IN KARLOVAC, TESLA FINISHED A FOUR-YEAR SCHOOL CURRICULUM IN THREE YEARS.
TESLA CAN PERFORM INTEGRAL CALCULUS IN HIS HEAD.
HE MUST BE CHEATING.

WARNED BY HIS FATHER NOT-TO RETURN TO GOSPIC BECAUSE OF A CHOLERA EPIDEMIC...

HE RETURNED ANYWAY...

AND PROMPTLY CAUGHT THE ILLNESS. BEDRIDDEN FOR NINE MONTHS, HE CAME NEAR DEATH MANY TIMES.

TESLA'S FATHER, WHO HAD LONG WANTED HIS SON TO JOIN THE CLERGY, MADE A PROMISE TO HIM.

IF YOU GET WELL, NIKO, I WILL SEND YOU TO THE BEST TECHNICAL INSTITUTION IN THE WORLD.
IT'S A DEAL, FATHER.

FIRST THERE WAS THE PROBLEM OF HOW TO AVOID BEING DRAFTED INTO THE AUSTRO-HUNGARIAN ARMY.
I'M FEELING BETTER.

A MAJOR WAR WAS BREAKING OUT AGAINST THE TURKS.
PACK WHAT YOU NEED AND GO AND HIDE IN THE HILLS.

FOR MOST OF THIS TERM I ROAMED THE MOUNTAINS LOADED WITH A HUNTER'S OUTFIT AND A BUNDLE OF BOOKS.

CONTACT WITH THE WAR WAS UNAVOIDABLE.

I HAVE SEEN MEN HUNG, BEATEN TO DEATH, SHOT, QUARTERED, STUCK ON A POINTED STICK...

HEADS CHOPPED OFF, AND CHILDREN ON A BAYONETTE, LIKE QUAILS 'EN BROCHE' AT DELMONICO'S.

IN 1875, HAVING GAINED A FELLOWSHIP FROM THE MILITARY FRONTIER AUTHORITY, TESLA WENT TO GRAZ, AUSTRIA, AND ENROLLED IN THE POLYTECHNIC THERE.

IN THAT FIRST YEAR, HE NEVER MISSED A LECTURE. HE EARNED THE HIGHEST GRADES, AND PASSED NINE EXAMS (TWICE WHAT WAS NEEDED).

THINGS DID NOT GO WELL IN THE THIRD YEAR. TESLA BECAME ADDICTED TO GAMBLING. IN ONE TERM HE LOST HIS ENTIRE ALLOWANCE.

DISCHARGED FROM THE POLYTECHNIC WITHOUT GRADUATING, HE CUT HIMSELF OFF FROM HIS FAMILY TO HIDE HIS FAILURE.

SEVERAL YEARS OF WANDERING THEN COMMENCED. FIRST SLOVENIA, WHERE HE WORKED AS AN ENGINEER. THEN TO GOSPIC, WHERE HE WAS ACCEPTED BACK INTO THE FAMILY.

HE CONTINUED HIS EDUCATION IN PRAGUE, BOHEMIA (NOW THE CZECH REPUBLIC).

IN 1881, TESLA WENT TO BUDAPEST, IN HUNGARY, WHERE HE BEGAN WORK AT THE NEWLY COMPLETED TELEPHONE EXCHANGE. HERE HE LEARNT HOW THE MOST MODERN INVENTIONS OF THE DAY WORKED.

THROUGHOUT THIS TIME HE WORKED ON A DESIGN FOR AN ELECTRICAL GENERATOR. HE INTENDED TO IMPROVE ON A MACHINE HE'D SEEN DEMONSTRATED IN GRAZ WHEN HE WAS A STUDENT.

THAT GENERATOR WAS POWERED BY DIRECT CURRENT (AN ELECTRIC CURRENT FLOWING IN ONE DIRECTION ONLY).

TESLA'S EVENTUAL DESIGN WOULD USE ALTERNATING CURRENT (AN ELECTRIC CURRENT THAT REVERSES DIRECTION MANY TIMES A SECOND).

A RING OF FOUR ELECTROMAGNETS WAS PLACED AROUND A FREELY ROTATING MOTOR PART (THE ROTOR).

ELECTRO MAGNETIC MOTOR.
PATENTED 1ST MAY 1888.

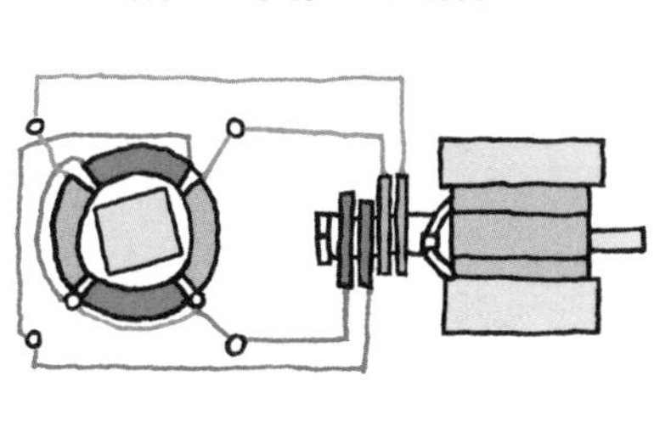

THE ELECTROMAGNETIC COILS, SHOWN HERE IN RED AND BLUE, WERE SWITCHED ON AND OFF ALTERNATELY.

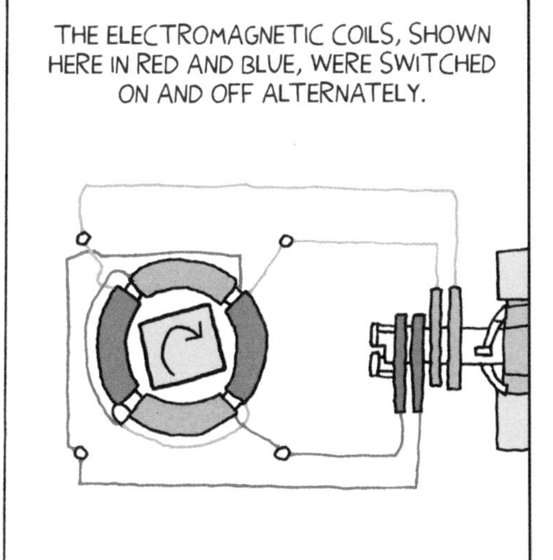

AS THE COILS WERE ENERGISED, THE MAGNETIC FIELD PRODUCED AN ELECTRIC CURRENT IN THE ROTOR.

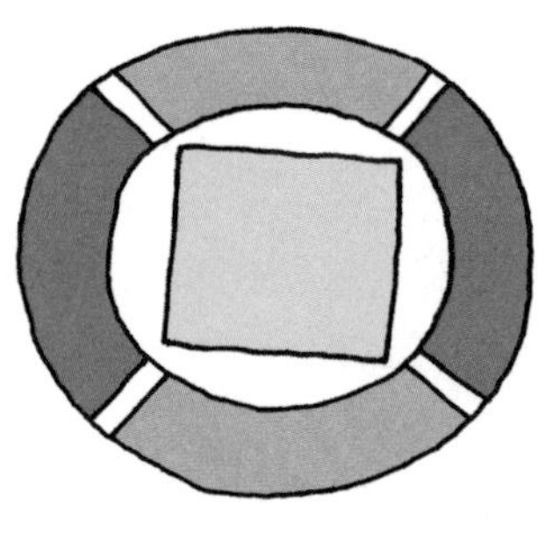

THIS CURRENT CREATED ITS OWN MAGNETIC FIELD. THE INTERACTION BETWEEN THE TWO FIELDS INDUCED THE ROTOR TO TURN.

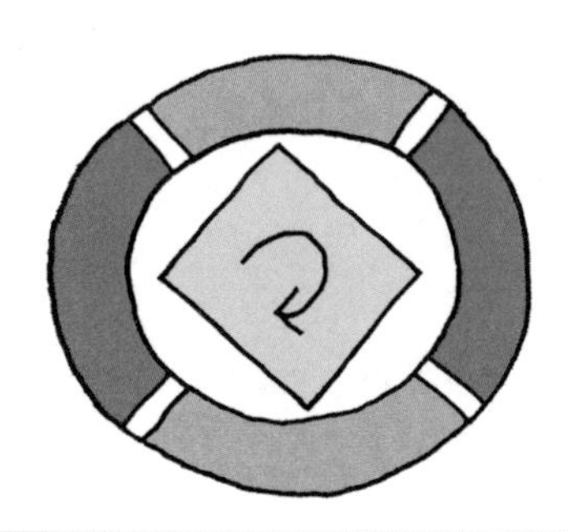

WAS TESLA THE FIRST TO CONCEIVE OF A ROTATING MAGNETIC FIELD? HE WAS NOT.

THE BRITISH INVENTOR, WALTER BAILEY, WAS THE FIRST TO DESIGN A ROTATING MAGNETIC FIELD, SIMILAR TO TESLA'S, IN 1879.

ALSO, QUITE INDEPENDENTLY, GALILEO FERRARIS DEVELOPED A VERSION OF THE MOTOR IN TURIN IN 1885.

TESLA DIDN'T DEMONSTRATE HIS DESIGN UNTIL 1887.

THOMAS EDISON WAS AN AMERICAN BUSINESSMAN AND INVENTOR. HE DEVELOPED MANY ELECTRICAL DEVICES, INCLUDING THE PHONOGRAPH, THE MOTION PICTURE CAMERA AND THE LIGHT BULB.

THE ERA OF GAS LIGHTING IN STREETS AND HOMES WAS ALMOST OVER. THE RACE TO BUILD A NEW LIGHTING INFRASTRUCTURE HAD BEGUN. IN 1882, THE EDISON ELECTRIC ILLUMINATING COMPANY ESTABLISHED THE FIRST INVESTOR-OWNED ELECTRIC POWER STATION, ON PEARL STREET, NEW YORK.

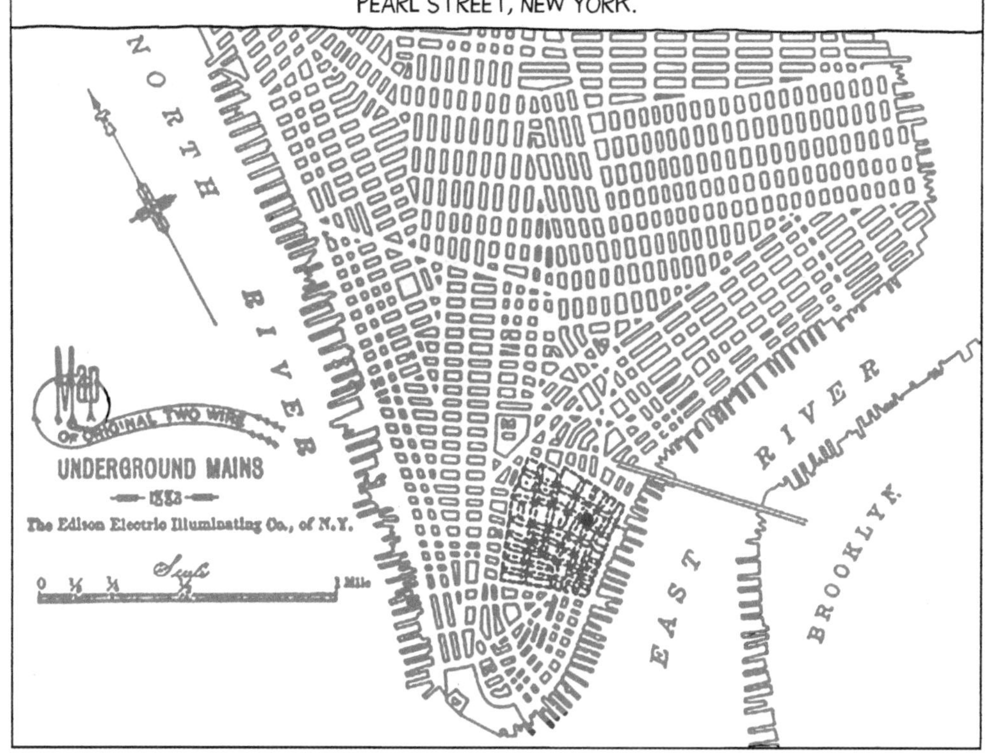


IN SEPTEMBER OF THAT YEAR, EDISON SWITCHED ON THE GENERATING STATION'S DISTRIBUTION SYSTEM, WHICH PROVIDED 110 VOLTS OF DIRECT CURRENT TO 59 CUSTOMERS IN LOWER MANHATTAN.

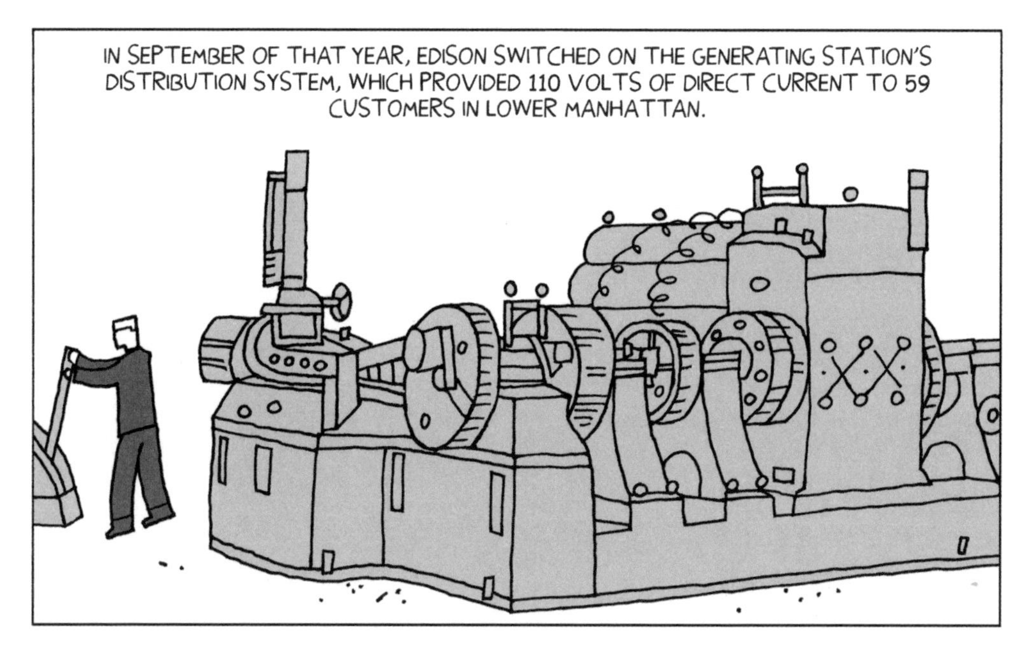

YET EDISON'S DIRECT CURRENT SYSTEM HAD A SERIOUS DRAWBACK. IT COULD NOT DELIVER ELECTRICITY MORE THAN ONE MILE FROM THE PLANT. HIS COMPETITORS, MAINLY GEORGE WESTINGHOUSE, FAVOURED ALTERNATING CURRENT (AC).

AC'S HIGH VOLTAGE COULD BE TRANSMITTED THROUGH THINNER AND CHEAPER WIRES THAN DC, AND ITS RANGE WAS MUCH GREATER. THE VOLTAGE WOULD THEN BE REDUCED OR STEPPED DOWN, FOR USE BY THE CUSTOMER.

EDISON WAS CRITICAL OF HIS COMPETITORS' USE OF HIGH VOLTAGE OVERHEAD CABLES, WHICH HE CONSIDERED DANGEROUS. HIS PREFERENCE WAS TO USE UNDERGROUND PIPES TO CARRY CABLING.

INTO THIS DEVELOPING 'WAR OF CURRENTS' WALKED NIKOLA TESLA.
HEH!

FOR EDISON, TESLA DID A RANGE OF WORK, WHICH INCLUDED FIXING BOTH INDOOR AND OUTDOOR INCANDESCENT LAMPS.

HE REDESIGNED 24 DIFFERENT TYPES OF MACHINES WHICH LATER BECAME STANDARD AND WHICH REPLACED THOSE EDISON USED. HE ALSO WORKED ON PATENTS FOR ARC LAMPS (A LAMP THAT PRODUCES LIGHT BY AN ELECTRIC ARC), REGULATORS (DESIGNED TO MAINTAIN A CONSTANT VOLTAGE LEVEL), DYNAMOS (MACHINES FOR CONVERTING MECHANICAL ENERGY INTO ELECTRICAL ENERGY), AND VOLTAGE CONVERTERS FOR DC APPARATUS.

ACCORDING TO TESLA, EDISON OFFERED HIM 50,000 DOLLARS FOR THE DESIGN WORK, YET WHEN THE TIME FOR PAYMENT ARRIVED HE CLAIMED TO HAVE BEEN JOKING.
WHEN YOU BECOME A FULL-FLEDGED AMERICAN, YOU WILL APPRECIATE AN AMERICAN JOKE.

EDISON OFFERED HIM A 10-DOLLAR RAISE OVER THE 18 DOLLARS HE WAS ALREADY BEING PAID, BUT TESLA REFUSED.

AFTER THIS, TESLA PATENTED A NUMBER OF REFINEMENTS TO THE DESIGN OF ARC LIGHTS AND ELECTRICITY GENERATORS.

HE FORMED A NEW COMPANY, THE TESLA ELECTRIC LIGHT MANUFACTURING COMPANY, WITH TWO NEW JERSEY BUSINESSMEN...

WHO EVENTUALLY FORCED HIM OUT.

NOW PENNILESS, TESLA EVEN LOST CONTROL OF HIS PATENTS, AS HE'D EXCHANGED THEM FOR STOCK IN THE BUSINESS. HE WAS LEFT...

TESLA RECOVERED FROM THIS SETBACK, IN PART THANKS TO GEORGE WESTINGHOUSE OF THE GEORGE WESTINGHOUSE ELECTRIC AND MANUFACTURING COMPANY.

WESTINGHOUSE WAS AN ENGINEER AND BUSINESSMAN WHOSE INITIAL SUCCESS HAD COME THOUGH INVENTING THE RAILWAY AIRBRAKE. HIS USE OF ALTERNATING CURRENT MADE HIM EDISON'S CHIEF COMPETITOR IN THE BUILDING OF AN ELECTRICITY INFRASTRUCTURE.

IN 1888 TESLA DEMONSTRATED HIS AC SYSTEM, INCLUDING HIS INDUCTION MOTOR, AT THE AMERICAN INSTITUTE OF ENGINEERS.

WHEN ENGINEERS WORKING FOR WESTINGHOUSE TOLD HIM OF THIS MOTOR HE WAS IMMEDIATELY INTERESTED.

WESTINGHOUSE HAD BEEN BUYING UP PATENTS FOR AC DEVICES WITH THE AIM OF BUILDING A COMPLETELY INTEGRATED AC SYSTEM.

HE'D CONSIDERED THE INDUCTION MOTOR DEVELOPED BY GALILEO FERRARIS, BUT IN THE END ACQUIRED TESLA'S DESIGN.

HOWEVER, EVEN WITH TESLA HIMSELF WORKING AT THE WESTINGHOUSE RESEARCH LABS IN PITTSBURGH...

IT WASN'T UNTIL 1893 THAT THE WESTINGHOUSE ENGINEER BENJAMIN G. LAMME DEVELOPED AN EFFICIENT VERSION OF THE INDUCTION MOTOR.

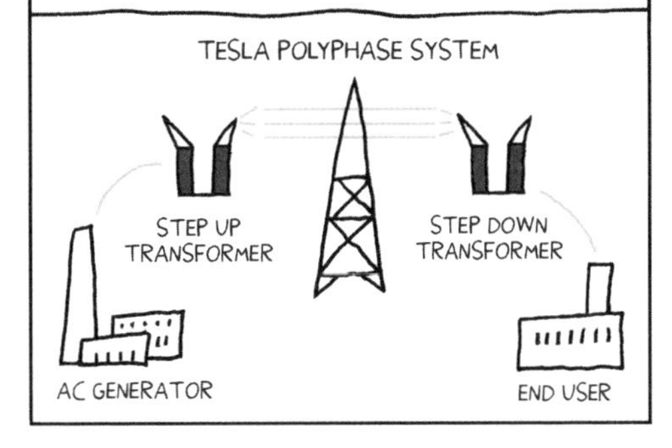


THROUGHOUT THIS TIME EDISON HAD BEEN ATTEMPTING TO DISCREDIT AC.

JUST AS CERTAIN AS DEATH, WESTINGHOUSE WILL KILL A CUSTOMER WITHIN SIX MONTHS AFTER HE PUTS IN A SYSTEM OF ANY SIZE.

EDISON'S CAMPAIGN TO DISCREDIT AC INCLUDED SPREADING DISINFORMATION ABOUT FATAL ACCIDENTS AND LOBBYING AGAINST THE USE OF AC IN STATE LEGISLATURES. HIS TECHNICIANS USED AC TO KILL ANIMALS: STRAY DOGS, CATS, BUT ALSO HORSES AND CATTLE, IN ORDER TO DEMONSTRATE ITS DANGERS.

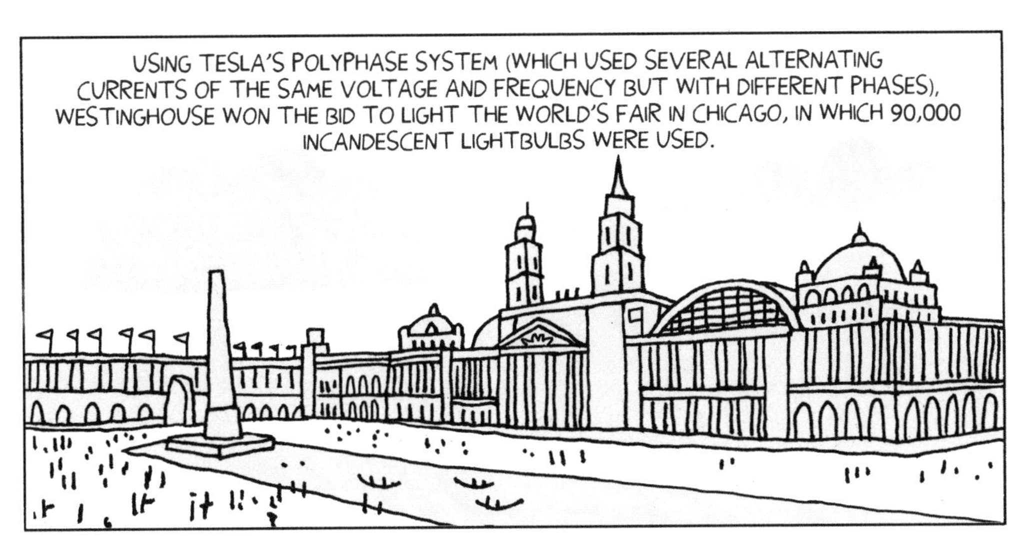
USING TESLA'S POLYPHASE SYSTEM (WHICH USED SEVERAL ALTERNATING CURRENTS OF THE SAME VOLTAGE AND FREQUENCY BUT WITH DIFFERENT PHASES), WESTINGHOUSE WON THE BID TO LIGHT THE WORLD'S FAIR IN CHICAGO, IN WHICH 90,000 INCANDESCENT LIGHTBULBS WERE USED.

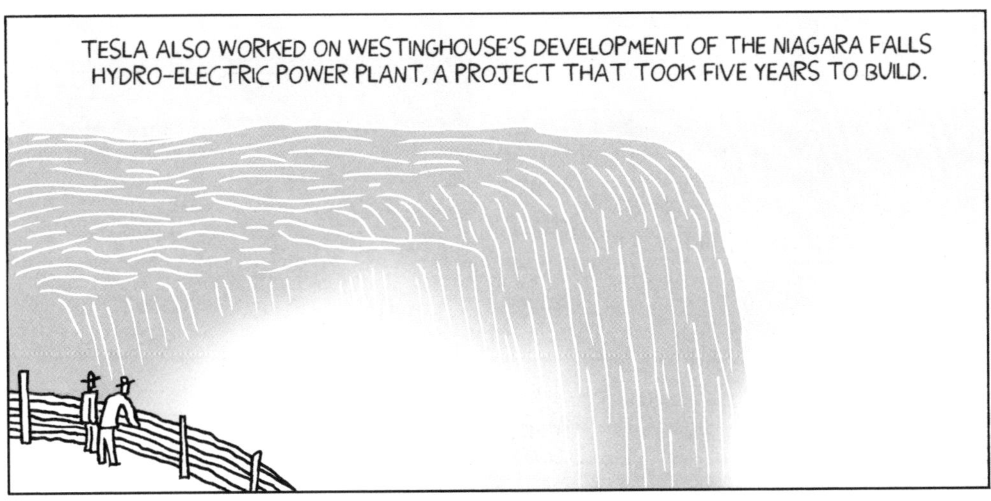
TESLA ALSO WORKED ON WESTINGHOUSE'S DEVELOPMENT OF THE NIAGARA FALLS HYDRO-ELECTRIC POWER PLANT, A PROJECT THAT TOOK FIVE YEARS TO BUILD.

THE TECHNICAL DIFFICULTIES WERE IMMENSE. THREE 5,000-HORSE-POWER ALTERNATORS WERE NEEDED – THE LARGEST EVER BUILT.

THE CURRENT FROM THE GENERATORS WAS STEPPED UP BY TRANSFORMERS TO 22,000 VOLTS, AND SENT OUT OVER LONG-DISTANCE LINES.

THE FIRST POWER REACHED BUFFALO ON 16TH NOVEMBER 1896.

WITHIN A FEW YEARS THE NUMBER OF GENERATORS AT NIAGARA HAD INCREASED TO TEN, AND POWER LINES STRETCHED 360 MILES TO NEW YORK.

BROADWAY BLAZED WITH LIGHTS. THE ELEVATED TRAINS AND SUBWAYS RAN ON AC.

THE WAR OF THE CURRENTS WAS OVER.

EDISON LOST CONTROL OF HIS COMPANY, WHICH WAS CONSOLIDATED INTO THE CONGLOMERATE GENERAL ELECTRIC...

WHICH THEN BEGAN CONVERTING TO AN AC DELIVERY SYSTEM.
BAH!

GENERAL ELECTRIC WOULD SOON CATCH UP WITH WESTINGHOUSE. A PATENT-SHARING AGREEMENT WAS SIGNED BETWEEN THE TWO COMPANIES.

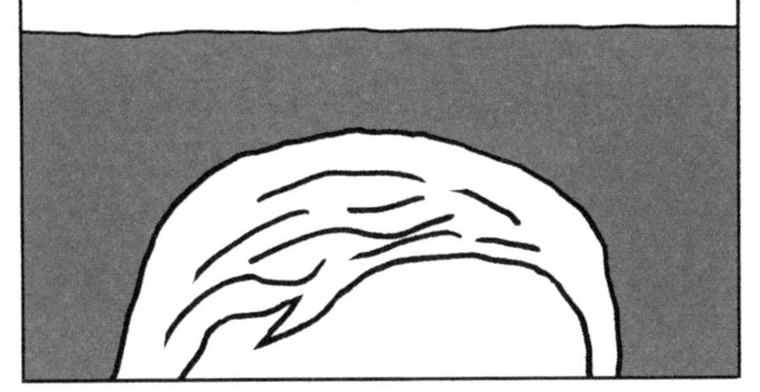

MEANWHILE VARIOUS INDEPENDENTS MERGED WITH WESTINGHOUSE...

TO FORM THE WESTINGHOUSE ELECTRIC AND MANUFACTURING COMPANY.

THE TESLA POLYPHASE SYSTEM PATENTS WERE THE MAIN ASSETS WESTINGHOUSE HAD TO OFFER, BUT HIS BANKERS DIDN'T WANT THE ONGOING ROYALTY DEAL WITH TESLA, BECAUSE OF THE EXPENSE.

TESLA WAS TALKED INTO SELLING HIS PATENTS FOR 216,000 DOLLARS. A MISTAKE THAT MAY HAVE COST HIM UP TO TWELVE MILLION.

POOR BUSINESS SENSE AND A SHORTAGE OF MONEY WERE RECURRING PROBLEMS FOR TESLA.

NOW HIS OWN BOSS, HE WAS ABLE TO FOLLOW HIS OWN INTERESTS.

THE RATE OF REVERSAL OF DIRECTION IN AC CURRENT IS KNOWN AS THE FREQUENCY. TESLA BEGAN TO EXPERIMENT WITH HIGH RATES OF FREQUENCY.

HE INVENTED A HIGH-FREQUENCY RESONANT TRANSFORMER CIRCUIT KNOWN AS THE TESLA COIL. THIS DEVICE WAS ABLE TO PRODUCE POWERFUL ELECTRIC FIELDS...

ENABLING HIM TO TRANSMIT ELECTRICITY THROUGH THE AIR.

WIRELESS POWER TRANSMISSION WAS TO BECOME A LIFELONG OBSESSION.

A DISCHARGE TUBE IS A TUBE FILLED WITH GAS THAT HAS ELECTRODES AT EITHER END. WHEN PLACED IN AN ELECTRIC FIELD, IONISATION IS INDUCED AND THE GAS WILL EMIT LIGHT. IONISATION IS THE PROCESS BY WHICH AN ATOM ACQUIRES A NEGATIVE OR POSITIVE CHARGE BY GAINING OR LOSING ELECTRONS.

IT WAS WHILE EXPERIMENTING WITH A DISCHARGE TUBE THAT TESLA NOTICED THAT NEARBY PHOTOGRAPHIC PLATES HAD BEEN DAMAGED.

BUT HE DID NOT PURSUE THE INVESTIGATION. IN 1895, WHEN WILHELM ROENTGEN PUBLISHED A PAPER ANNOUNCING THE DISCOVERY OF X-RAYS, TESLA REALISED WHAT HE HAD MISSED.

HE WENT ON TO DESIGN HIS OWN X-RAY EQUIPMENT AND TOOK MANY OF THE FIRST X-RAYS OF THE HUMAN BODY.

THAT SAME YEAR A FIRE DESTROYED TESLA'S NEW YORK LABORATORY, WHICH OCCUPIED THE ENTIRE FOURTH FLOOR OF A SIX-STOREY BUILDING.

GONE WERE HUNDREDS OF INVENTION MODELS, PLANS, NOTES, DATA, TOOLS AND PHOTOGRAPHS.

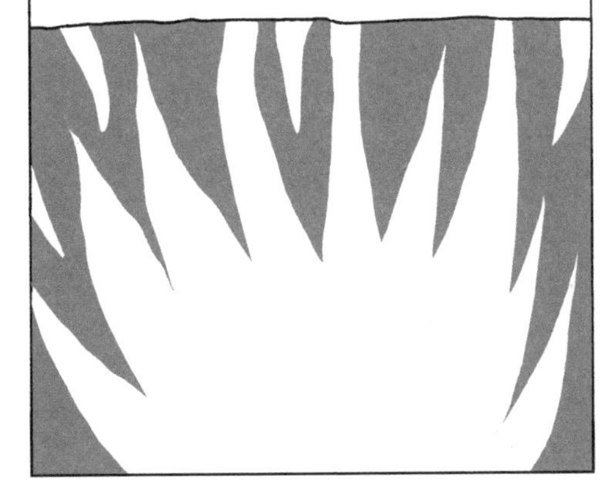

THE FIRE HAD STARTED, NOT IN THE LAB, BUT IN A DRY CLEANING ESTABLISHMENT ON THE FLOOR BELOW.

TESLA HAD NOT INSURED THE PROPERTY.
I AM IN TOO MUCH GRIEF TO SPEAK. WHAT CAN I SAY?

THE WESTINGHOUSE COMPANY BILLED TESLA FOR THE COST OF THE MACHINERY THEY HAD LOANED HIM. THEY ALSO CHARGED HIM FOR NEW EQUIPMENT.
BILL

FOR A FEW WEEKS, UNTIL HE COULD FIND A NEW LAB, TESLA USED EDISON'S WORKSHOP IN NEW JERSEY...

A CURIOUS THING TO HAPPEN IF THE RIVALRY BETWEEN THE TWO MEN WAS AS INTENSE AS THE PRESS AT THE TIME DESCRIBED.

WITH A 40,000-DOLLAR GRANT FROM A BACKER, TESLA EQUIPPED A NEW LABORATORY IN GREENWICH VILLAGE.

ALTHOUGH TESLA WAS A FASTIDIOUS MAN, ALWAYS ELEGANTLY DRESSED, HE DID NOT TAKE AS MUCH CARE OF HIS HEALTH AS HE DID HIS APPEARANCE. HE OFTEN OVERWORKED TO THE POINT OF ILLNESS.

THIS WAS A CONCERN OF HIS CLOSEST FRIENDS, ROBERT UNDERWOOD JOHNSON, POET AND EDITOR OF CENTURY MAGAZINE, AND HIS WIFE, KATHARINE.

KATHARINE UNDERWOOD'S INTEREST IN TESLA OFTEN TEETERED ON THE EDGE OF SOMETHING MORE.

WILL YOU COME TO SEE ME TOMORROW EVENING, AND WILL YOU TRY AND COME A LITTLE EARLIER?

TESLA WAS A LIFELONG BACHELOR. HIS RELATIONSHIPS WITH BOTH SEXES APPEAR TO HAVE BEEN PLATONIC.

I DESTROYED MY SEXUALITY AT THE AGE OF 33 BECAUSE A CERTAIN FRENCH ACTRESS KEPT COMING TO ME, MAKING IT IMPOSSIBLE FOR ME TO CONCENTRATE.

THE FRENCH ACTRESS MAY HAVE BEEN SARAH BERNHARDT. WHAT DID TESLA MEAN BY HIS COMMENT THAT HE'D 'DESTROYED HIS SEXUALITY'? WAS THIS PHYSICAL DAMAGE, OR SOME MENTAL DISCIPLINE HE BELIEVED HE HAD ATTAINED?

TESLA WAS SO PHOBIC THAT HE COULD NOT SHAKE HANDS OR TOUCH A STRAND OF HAIR. GERMS WERE A CONSTANT FEAR. HE'D OBSESSIVELY WASH HIS HANDS, SOMETIMES TWO OR THREE TIMES DURING A MEAL. INTIMACY MAY HAVE BEEN IMPOSSIBLE FOR HIM.

I DO NOT THINK YOU CAN NAME MANY GREAT INVENTORS THAT HAVE BEEN MARRIED MEN. IT IS A PITY, TOO, FOR SOMETIMES I FEEL SO LONELY.

IN 1898 TESLA DEMONSTRATED A RADIO-CONTROLLED BOAT DURING AN ELECTRICAL EXHIBITION HELD AT MADISON SQUARE GARDEN.

THIS WAS AT THE HEIGHT OF THE SPANISH-AMERICAN WAR. TESLA HOPED TO SELL THE IDEA TO THE MILITARY, BUT THEY SHOWED LITTLE INTEREST.

BY THE NEXT YEAR HIS NEW YORK LAB HAD BECOME TOO SMALL FOR THE HIGH-VOLTAGE AND HIGH-FREQUENCY EXPERIMENTS HE WAS CONDUCTING. HE MOVED TO COLORADO.

THE EL PASO POWER COMPANY OF COLORADO SPRINGS AGREED TO SUPPLY THE ELECTRICITY AND LAND FOR TESLA'S NEW RESEARCH STATION. WEALTHY BACKERS, INCLUDING JOHN JACOB ASTOR, SUPPLIED THE MONEY.

HERE HE BUILT A SPARK GAP TRANSMITTER, THE LARGEST EVER BUILT, WITH THE AIM OF BROADCASTING POWER WIRELESSLY OVER LONG DISTANCES.

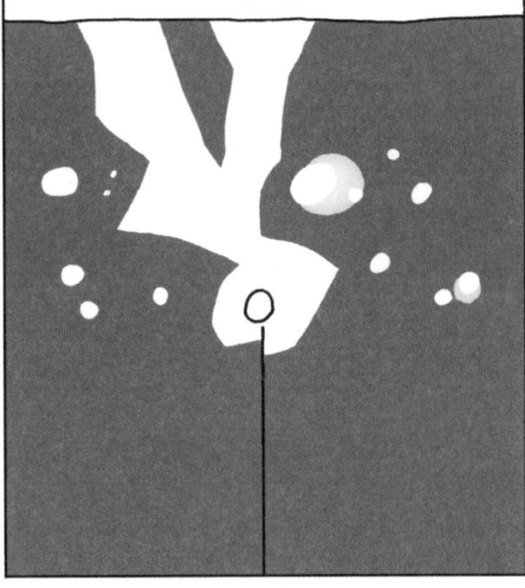
WHEN THE MACHINE WAS FIRST USED IT SHOT LIGHTNING FROM ITS TOWER 135 FEET INTO THE AIR. THE LOCAL PAPER REPORTED THAT THE BOOM COULD BE HEARD IN CRIPPLE CREEK, 20 MILES AWAY.

THE TEST WAS SHORT-LIVED. THE TRANSMITTER OVERLOADED COLORADO SPRINGS' GENERATOR, CAUSING A FIRE...

PLUNGING THE TOWN INTO DARKNESS.

DURING HIS TIME IN COLORADO, TESLA CLAIMED TO HAVE RECEIVED UNUSUAL SIGNALS FROM SPACE. A CLEAR, REPEATED ONE, TWO, THREE...

WHICH HE BELIEVED WERE COMMUNICATIONS FROM ANOTHER PLANET, PERHAPS MARS? A STATEMENT HE WAS RIDICULED FOR.

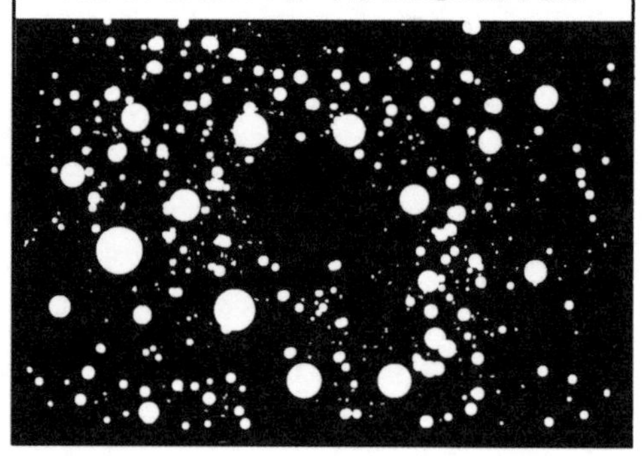

TESLA MAY HAVE INTERCEPTED ONE OF GUGLIELMO MARCONI'S BROADCASTS.

MARCONII WAS AN ITALIAN NOBLEMAN, INVENTOR AND ELECTRICAL ENGINEER, WHO AT THIS TIME WAS WORKING ON LONG-DISTANCE TRANSMISSIONS.

WHETHER TESLA HEARD MARCONI, ANOTHER RADIO PIONEER, OR WAS THE FIRST TO PICK UP NATURAL COSMIC RADIO SIGNALS, IS ANOTHER OF THE MYSTERIES THAT SURROUND HIM.

IN THE END, THE COLORADO POWER COMPANY SUED TESLA FOR THE ELECTRICITY USED, AND THE LAB WAS TORN DOWN AND USED FOR LUMBER.

TESLA RETURNED TO NEW YORK WHERE HE MOVED INTO THE WALDORF ASTORIA.

KNOWING HE WOULD GET NO MORE MONEY FROM ASTOR, WHO WAS BARELY ANSWERING HIS LETTERS, TESLA LOOKED TO J.P. MORGAN FOR FURTHER INVESTMENT.

JOHN PIERPOINT MORGAN WAS AN AMERICAN FINANCIER AND BANKER WHO DOMINATED CORPORATE FINANCE IN HIS ERA.

HE HAD ARRANGED THE MERGER OF EDISON GENERAL ELECTRIC AND THE THOMAS HOUSTON ELECTRIC COMPANY TO FORM GENERAL ELECTRIC.

HE WAS INSTRUMENTAL IN THE CREATION OF THE UNITED STATES STEEL CORPORATION.

I WILL PAY YOU 150,000 DOLLARS FOR THE BUILDING OF YOUR TRANSMISSION TOWER AND TAKE A FIFTY-ONE PER CENT STAKE ON THE COMPANY'S STOCK AND YOU FORTY-NINE.
IT'S A DEAL.

ALL THAT MORGAN WANTED WAS TO SIGNAL OCEAN LINERS AND SEND MORSE CODE TO EUROPE.

BUT TESLA HAD MORE AMBITIOUS PLANS. HIS AIM WAS TO TRANSMIT INDUSTRIAL LEVELS OF ELECTRICAL POWER WITHOUT WIRES...

AND TO DO THIS ON A GLOBAL SCALE.

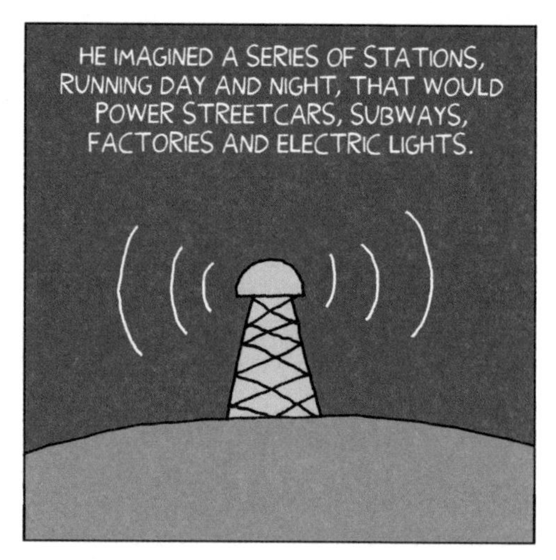
HE IMAGINED A SERIES OF STATIONS, RUNNING DAY AND NIGHT, THAT WOULD POWER STREETCARS, SUBWAYS, FACTORIES AND ELECTRIC LIGHTS.

EACH PLANT WOULD BE LOCATED NEAR A MAJOR POPULATION CENTRE.

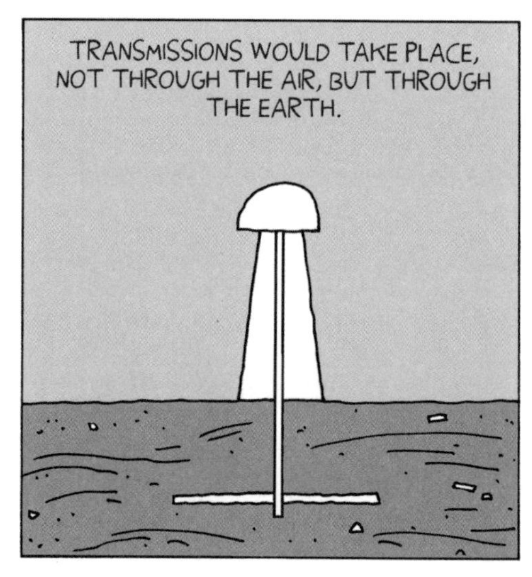
TRANSMISSIONS WOULD TAKE PLACE, NOT THROUGH THE AIR, BUT THROUGH THE EARTH.

THE TOWERS WOULD ALSO TRANSMIT WIRELESS TELEGRAPH AND TELEPHONE SERVICES WORLDWIDE.

'TESLA DREAMS OF THE DAY WHEN A MAN MAY DICTATE HIS ORDERS IN SAN FRANCISCO...

'AND THEY MAY BE SET UP AND PRINTED DIRECTLY IN NEW YORK.

'OF A DAY WHEN NOT MERELY PHOTOGRAPHS WILL BE TRANSMITTED BY WIRELESS, BUT WHEN ONE CAN ACTUALLY SEE, OVER THE WIRELESS, THE FRIEND TO WHOM HE IS TALKING IN SOME DISTANT PART OF THE EARTH.'

'TESLA AND HIS WIRELESS AGE'. *POPULAR ELECTRONICS*, 1911.

SUCH IDEAS MAY SEEM ASTONISHINGLY PRESCIENT, AND THEY ARE, BUT DREAMS ARE ONE THING AND REALITY IS ANOTHER. NEVER SHORT ON SELF-CONFIDENCE, TESLA'S VANITY AND TENDENCY TOWARDS GRANDIOSITY CAUSED HIM TO REACH FAR BEYOND HIS ABILITY TO DELIVER.

IN 1900, TESLA BEGAN PLANNING THE FIRST OF HIS TOWERS, TO BE BUILT IN SHOREHAM, LONG ISLAND, NEW YORK. BUT, BEFORE CONSTRUCTION COULD EVEN START ON WHAT WOULD BE KNOWN AS THE WARDENCLYFFE TOWER, THE NEWS BROKE THAT MARCONI HAD SUCCESSFULLY TRANSMITTED THE LETTER 'S' FROM ENGLAND TO NEWFOUNDLAND, BEATING TESLA IN THE RACE TO WIRELESSLY SEND INFORMATION ACROSS THE ATLANTIC.

TESLA PRESSED ON ANYWAY. AFTER NINE MONTHS THE TOWER REACHED ITS FULL HEIGHT OF 187 FEET. BUT, OF COURSE, IT WAS TOO LATE. MARCONI WOULD FOREVER BE KNOWN AS THE FATHER OF RADIO AND TESLA A MERE FOOTNOTE. NOT ONLY WOULD THE ITALIAN REAP THE COMMERCIAL BENEFITS OF HIS SUCCESS, BUT HE WOULD ALSO WIN THE NOBEL PRIZE, SHARING IT WITH FELLOW RADIO PIONEER, KARL BRAUN.

TELSA'S WORLD WIRELESS NETWORK HAD FAILED. HAD HE CONCENTRATED ON DEVELOPING RADIO, THEN PERHAPS HE WOULD HAVE BEATEN MARCONI. BUT HE WAS OBSESSED WITH BROADCAST POWER TO THE EXCLUSION OF ALL ELSE.

IT WAS AN UNWORKABLE DREAM. CALCULATIONS SHOULD HAVE TOLD HIM THAT YOU CAN WIRELESSLY TRANSMIT ELECTRICAL POWER, BUT NOT IN NEARLY ENOUGH QUANTITIES TO RUN HOMES, BUSINESSES AND CITY INFRASTRUCTURE.

THE POWER IS DILUTED, BECAUSE SPACE IS BIG. IT IS SIMPLY MORE EFFICIENT TO USE CABLES.

OVER THE NEXT FIVE YEARS, TESLA WROTE MORE THAN 50 INCREASINGLY DESPERATE LETTERS TO J.P. MORGAN, PLEADING FOR MORE FUNDING, BUT THE FINANCIER DECLINED ANY FURTHER INVOLVEMENT.

AT THE AGE OF 50, TESLA WAS BROKE. HE HAD NUMEROUS CREDITORS. HIS BILL AT THE WALDORF ASTORIA ALONE CAME TO 20,000 DOLLARS, WHICH HE COULD NOT PAY.

AS A RESULT OF THIS, GEORGE C. BOLDT, THE HOTEL'S PROPRIETOR, TOOK OWNERSHIP OF WARDENCLYFFE AND THE TOWER WAS DEMOLISHED.

TESLA'S DREAM WAS OVER. AS EVER, HIS NEMESIS HAD NOT BEEN EDISON, MARCONI OR ANY OTHER RIVAL...

BUT HIS OWN UNWORLDLINESS AND POOR BUSINESS SENSE.

IN 1917, THE AMERICAN INSTITUTE OF ELECTRICAL ENGINEERS GAVE TESLA THE EDISON AWARD – THE HIGHEST HONOUR THEY HAD. THIS HE ACCEPTED, AFTER FIRST REFUSING...

BUT IT MUST HAVE BEEN A HOLLOW EXPERIENCE.

TESLA'S DECLINE THEN BEGAN. HE DID INVENT A BLADELESS TURBINE WHICH HE HOPED WOULD REVIVE HIS FORTUNES...

BUT THE MACHINE PROVED IMPRACTICAL USING THE MATERIALS OF THE TIME, AND NO MORE EFFICIENT THAN THE TURBINES ALREADY IN USE.

NEAR THE END OF HIS LIFE, TESLA OFTEN WALKED TO THE PARK, WHERE HE WOULD FEED THE PIGEONS.

HE SPENT OVER 2,000 DOLLARS ON AN INJURED PIGEON, INCLUDING BUILDING A DEVICE THAT SUPPORTED HER, SO HER BONES COULD HEAL.

I HAVE BEEN FEEDING PIGEONS, THOUSANDS OF THEM, FOR YEARS, BUT THERE WAS ONE, A BEAUTIFUL BIRD, PURE WHITE WITH LIGHT GREY TIPS ON ITS WINGS.

I HAD ONLY TO WISH AND CALL HER AND SHE WOULD COME FLYING TO ME. I LOVED THAT PIGEON AS A MAN LOVES A WOMAN AND SHE LOVED ME. AS LONG AS I HAD HER, THERE WAS PURPOSE TO MY LIFE.

IT'S CURIOUS THAT HE BONDED WITH PIGEONS WHEN HE WAS SUCH A GERMOPHOBE. IT SAYS MUCH ABOUT THE DEPTHS OF HIS LONELINESS.

AS HE AGED, TESLA LIVED ON AN EVER-RECEDING SUBSISTENCE DIET. FIRST HE GAVE UP MEAT...

MOVING TO FISH, THEN VEGETABLES, AND FINALLY A CONCOCTION HE CALLED FACTOR-ACTIVE...

A HEALTH DRINK COMPOSED OF A DOZEN VEGETABLES, INCLUDING WHITE LEEK, CABBAGE, CAULIFLOWER AND LETTUCE.

HE'D ALWAYS BEEN A SLIM MAN, BUT HE BECAME THINNER AND THINNER.

MUCH HAS BEEN WRITTEN ABOUT THE 'DEATH RAY' TESLA IS SAID TO HAVE INVENTED IN HIS LATTER YEARS.

ON THE INVENTOR'S 78TH BIRTHDAY IN 1934 HE MADE THE CLAIM, REPORTED IN THE *NEW YORK HERALD TRIBUNE,* THAT HE'D DESIGNED A WEAPON CAPABLE OF...

PRODUCING A CONCENTRATED BEAM OF SUB-MICROSCOPIC PARTICLES, FLYING AT VELOCITIES CLOSE TO THE SPEED OF LIGHT...

THAT WOULD DROP AN ARMY IN ITS TRACKS...

AND BRING DOWN SQUADRONS OF AEROPLANES FROM 250 MILES DISTANT.

BUT, AS WE HAVE SEEN, TESLA WAS A MAN PRONE TO EXAGGERATION, WHO OFTEN LACKED THE ABILITY TO DELIVER PROMISED RESULTS.

NO SUCH WEAPON EVER APPEARED, DESPITE HIS CLAIM THAT HE WAS NEGOTIATING WITH SEVERAL GOVERNMENTS.

IN 1934, THE WESTINGHOUSE ELECTRIC AND MANUFACTURING COMPANY BEGAN PAYING TESLA 125 DOLLARS A MONTH, PLUS HIS RENT AT THE NEW YORKER HOTEL. HE WOULD LIVE THERE FOR THE REST OF HIS LIFE. WHETHER THE WESTINGHOUSE COMPANY DID THIS BECAUSE THEY WERE WORRIED ABOUT BAD PUBLICITY SURROUNDING THE POVERTY OF THEIR ONE-TIME STAR INVENTOR, OR WHETHER IT WAS SOME FORM OF FINANCIAL SETTLEMENT, ISN'T CLEAR.

ONE EVENING, AS TESLA WAS LEAVING THE HOTEL TO FEED THE PIGEONS, HE WAS KNOCKED DOWN BY A TAXI CAB.

HE REFUSED TO SEE A DOCTOR – AN ALMOST LIFELONG CUSTOM. HIS BACK WAS SEVERELY WRENCHED AND HE'D CRACKED THREE RIBS. HE WAS TO STAY IN BED, ON AND OFF, FOR SIX MONTHS.

TESLA DIED ALONE IN HIS ROOM AT THE NEW YORKER HOTEL, ON 8TH JANUARY 1943. HE WAS 86 YEARS OLD.

THE CAUSE OF DEATH WAS CORONARY THROMBOSIS.

TWO DAYS LATER, ALL OF TESLA'S BELONGINGS WERE SEIZED BY THE OFFICE OF ALIEN PROPERTY, EVEN THOUGH THE INVENTOR HAD BEEN AN AMERICAN CITIZEN SINCE 1891.

THE MAN IN CHARGE OF ANALYSING TESLA'S NOTES AND DOCUMENTS WAS JOHN G. TRUMP (UNCLE OF THE RATHER BETTER KNOWN DONALD).

JOHN GEORGE TRUMP WAS NO SCIENTIFIC ILLITERATE. HE WAS A PHYSICS PROFESSOR AT THE MASSACHUSETTS INSTITUTE OF TECHNOLOGY.

AT THIS TIME HE WAS WORKING AS A TECHNICAL AIDE FOR THE NATIONAL DEFENCE RESEARCH COMMITTEE.

TESLA'S EFFECTS FILLED 88 TRUNKS. IT TOOK TRUMP AND HIS ASSISTANT TWO DAYS TO ASSESS ALL THE MATERIAL.

HE RETURNED TO WASHINGTON TO GIVE HIS REPORT. HIS CONCLUSION WAS THAT THE LAST FIFTEEN YEARS OF TESLA'S LIFE HAD BEEN...

PRIMARILY OF A SPECULATIVE AND PHILOSOPHICAL AND SOMEWHAT PROMOTIONAL NATURE.

IT IS MY OPINION THAT THE TESLA PAPERS CONTAIN NOTHING OF VALUE FOR THE WAR EFFORT...

AND NOTHING WHICH WOULD BE HELPFUL TO THE ENEMY, IF IT FELL INTO ENEMY HANDS.

ARE YOU QUITE SURE OF THIS, DR TRUMP?
I'M WILLING TO STAKE MY PROFESSIONAL REPUTATION ON IT.

TODAY, TESLA'S ASHES ARE IN A GOLD-PLATED SPHERE DISPLAYED ON A MARBLE PEDESTAL IN THE NIKOLA TESLA MUSEUM, BELGRADE, SERBIA.
P

CONTRARY TO POPULAR OPINION, TESLA WAS NOT A SCIENTIFIC WIZARD, UNJUSTLY NEGLECTED, WHO HAD HIS GREAT WORK STOLEN BY OTHERS. HE MADE KEY INNOVATIONS TO AC THAT HELPED USHER IN THE MODERN WORLD, BUT, EVEN THEN, HE WAS ONLY ONE OF MANY WHO WORKED IN THIS FIELD. HIS LIFE IS OF INTEREST TO US AS MUCH FOR HIS FAILURES AND HIS CONNECTIONS WITH OTHER GIANTS OF THE ERA - EDISON, WESTINGHOUSE, J.P. MORGAN AND MARCONI - AS FOR HIS SUCCESSES. HE WAS ALL TOO HUMAN, WITH FLAWS AND IDIOSYNCRASIES. WE SHOULD APPRECIATE THE MAN, NOT THE MYTH.
END

ALFRED
WEGENER

ALFRED LOTHAR WEGENER WAS BORN IN BERLIN IN 1880.

HE WAS THE FIFTH AND YOUNGEST CHILD OF RICHARD AND ANNA WEGENER. ALFRED'S FATHER WAS A MINISTER AND THEOLOGIAN, WHO, WITH HIS WIFE, RAN AN ORPHANAGE.

THIS WAS NO DISMAL DICKENSIAN INSTITUTION, BUT A PRIVATE SCHOOL FOR THE SONS OF CIVIL SERVANTS, CLERGYMEN AND TEACHERS.

IN 1886 THE WEGENER FAMILY BOUGHT A SECOND HOME, NORTH OF BERLIN, OUTSIDE RHEINSBERG.

THIS WAS TO BE THEIR HOLIDAY HOME AND SUMMER RESIDENCE. AN ESCAPE FROM THE URBAN, TREELESS WORLD OF THE CITY.

THE WEGENERS HAD FIVE CHILDREN, BUT THE YOUNGEST, ALFRED, AND HIS OLDER BROTHER, KURT, FORMED A PARTICULARLY STRONG BOND...

ONE BASED ON PHYSICAL ACTIVITY AND LOVE OF THE OUTDOORS.

THE BOYS LIKED TO ROW ON THE NEARBY LAKE SYSTEM, DRAWING MAPS, AND MAKING DEPTH SOUNDINGS.

THEY WERE NATURAL EXPLORERS. WITHIN A FEW YEARS THEY HAD GONE EVERYWHERE WITHIN A SIX-MILE RADIUS OF THE HOUSE.

FROM THEIR EARLIEST YEARS THE BROTHERS WERE EXPERIMENTERS AND EXPLORERS. IN 1897, THE SUMMER BEFORE KURT WENT OFF TO UNIVERSITY, THEY MADE A NUMBER OF LONG VOYAGES VIA CANAL TO LAKE DESTINATIONS 30 MILES AWAY.

THEY RETURNED HOME WITH THEIR FACES SO BADLY SWOLLEN FROM SUNBURN THAT THEY WERE UNRECOGNISABLE.

AT SUMMER'S END, KURT LEFT TO STUDY MATHEMATICS, ENGINEERING MECHANICS AND ADVANCED SURVEYING.

WHEN ALFRED WENT TO UNIVERSITY, HE CHOSE AS HIS SUBJECTS PHYSICS, ASTRONOMY AND METEOROLOGY.

AT THIS TIME, METEOROLOGY, THE STUDY OF THE WEATHER, WAS NOT SEPARATE FROM ASTRONOMY, BUT AN ESSENTIAL PART OF THE SUBJECT.

AN UNDERSTANDING OF WEATHER SYSTEMS WAS NECESSARY FOR ASTRONOMERS, AS ATMOSPHERIC CONDITIONS HAD TO BE CONSIDERED WHEN MAKING ANY OBSERVATION OF CELESTIAL OBJECTS.

ALTHOUGH WEGENER STARTED HIS EDUCATION AIMING TO BE AN ASTRONOMER, METEOROLOGY WAS TO BECOME HIS MAIN INTEREST.

IN ASTRONOMY EVERYTHING HAS ESSENTIALLY BEEN DONE. ONLY AN UNUSUAL TALENT FOR MATHEMATICS TOGETHER WITH SPECIALISED INSTALLATIONS AT OBSERVATORIES CAN LEAD TO NEW DISCOVERIES...

AND BESIDES, ASTRONOMY OFFERS NO OPPORTUNITIES FOR PHYSICAL ACTIVITY.

THIS LAST POINT WAS IMPORTANT TO BOTH WEGENERS. IN 1901 THEY TOOK A SUMMER SEMESTER TOGETHER AT THE UNIVERSITY OF INNSBRUCK IN AUSTRIA, TAKING FIELD GEOLOGY AND BIOLOGY. WHEN THE TERM ENDED THEY HEADED FOR THE ALPS, SPENDING THE REST OF THE SUMMER CLIMBING IN THE ZILLERTAL REGION TO THE EAST OF THE BRENNER PASS.

FOR THE TWO BROTHERS A LIFE IN A DULL OFFICE JOB OR TEACHING POST WAS NEVER GOING TO BE ACCEPTABLE. ONLY A FUTURE WHICH COMBINED SCIENCE, GEOGRAPHICAL EXPLORATION AND PHYSICAL ACTIVITY WAS GOING TO SATISFY THEM.

AFTER THIER ALPINE ADVENTURE, THE WEGENERS THEN COMPLETED THEIR COMPULSORY ONE-YEAR MILITARY SERVICE AS OFFICER CADETS IN THE GRENADIER GUARDS.

ALFRED WOULD REMAIN A LIEUTENANT IN THE RESERVES AFTER HE RETURNED TO FULL-TIME EDUCATION.

HE BEGAN A ROTATION AS A STUDENT ASSISTANT AT THE BERLIN OBSERVATORY.

AMONG HIS TUTORS AT THE UNIVERSITY WAS MAX PLANCK, THE PHYSICIST FROM WHOM HE TOOK LECTURES IN THERMODYNAMICS AND THERMOCHEMISTRY.

THE NOBEL PRIZE WINNER HAD A PROFOUND INFLUENCE ON WEGENER. FROM PLANCK, HE LEARNED TO PAY CLOSE ATTENTION TO THE FACTS OF A PHENOMENON: ITS TEMPERATURE, PRESSURE, MASS, AND VOLUME.

WHY A PROCESS WAS HAPPENING COULD BE TREATED AS A SEPARATE HYPOTHESIS. A THEORY COULD BE COMPLETE WITHOUT A CAUSE BEING FOUND. THIS ALLOWED ALFRED SOME FLEXIBILITY IN HIS THINKING, WHICH BECAME IMPORTANT IN HIS LATER WORK.

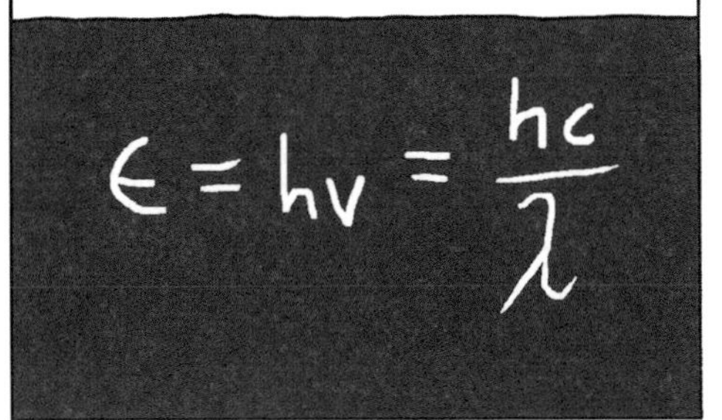

IN 1905, HAVING GAINED HIS DOCTORATE IN ASTRONOMY, WEGENER BECAME AN ASSISTANT AT THE AERONAUTISCHE OBSERVATORIUM, LINDENBERG, WHERE HIS BROTHER KURT ALREADY HELD A POSITION. THIS WAS THE MAIN GERMAN CENTRE FOR AEROLOGY, WHEREBY THE ATMOSPHERE WAS STUDIED USING BALLOONS AND KITES TO SEND METEOROLOGICAL INSTRUMENTS SEVERAL MILES INTO THE AIR.

IN 1903, WILBUR AND ORVILLE WRIGHT HAD SUCCESSFULLY MADE BRIEF FLIGHTS WITH THE FIRST POWERED AIRCRAFT, BUT IT WOULD BE YEARS BEFORE AEROPLANES WOULD MANAGE SUSTAINED JOURNEYS AT THE HEIGHTS NECESSARY TO STUDY THE ATMOSPHERE. UNTIL THEN, MANNED BALLOONS WERE IN COMMON USE TO GATHER DATA, AND THIS WAS NATURALLY OF GREAT INTEREST TO THE ADVENTUROUS WEGENERS.

IN 1906, THE WEGENER BROTHERS TOOK PART IN THE GORDON BENNETT INTERNATIONAL BALLOON COMPETITION, REPRESENTING LINDENBERG AND GERMANY. TEAMS OF AERONAUTS FROM ALL OVER EUROPE ENTERED THIS EVENT, WHICH WAS SPONSORED BY JAMES GORDON BENNETT, THE MILLIONAIRE SPORTSMAN AND OWNER OF THE *NEW YORK HERALD* NEWSPAPER.

FOR THE WEGENERS, THIS WAS NOT JUST A SPORTING OCCASION. THEY PLANNED TO EXPLOIT THE NIGHT FLIGHT TO PRACTISE NAVIGATION BY STAR SIGHTINGS. KURT, WHO WAS THE MORE EXPERIENCED BALLOONIST, TOOK CHARGE, WHILE ALFRED ACTED AS NAVIGATOR.

THAT MORNING THE BALLOON TOOK OFF RAPIDLY AND FLEW NORTHWEST, DRIFTING OVER MANY OF THE WOODS, LAKES AND RIVERS WHERE THE WEGENERS HAD SPENT THEIR CHILDHOOD SUMMERS.

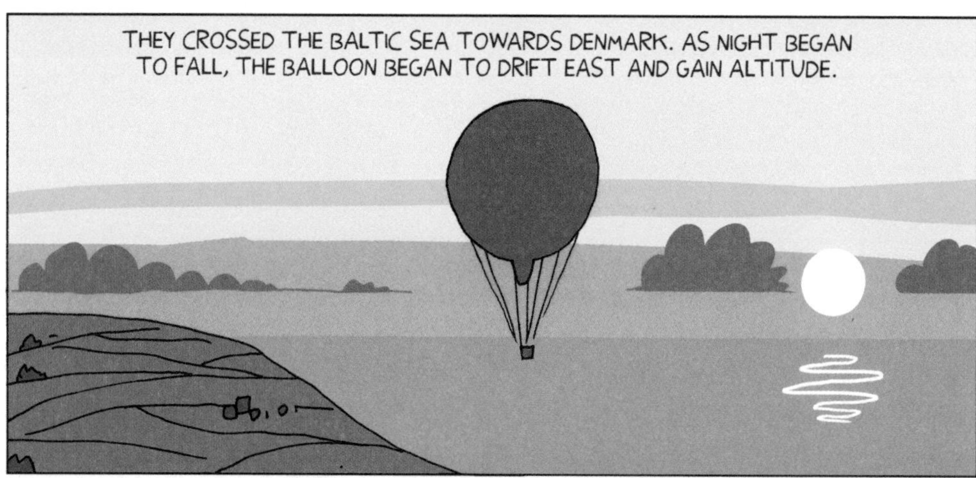
THEY CROSSED THE BALTIC SEA TOWARDS DENMARK. AS NIGHT BEGAN TO FALL, THE BALLOON BEGAN TO DRIFT EAST AND GAIN ALTITUDE.

THE TEMPERATURE DROPPED SHARPLY. IT WAS ONLY THEN THAT KURT AND ALFRED REALISED THEY HAD LEFT THEIR OVERCOATS BEHIND.

THE NEXT MORNING THEY WERE BOTH NEARLY FROZEN. THE BALLOON HAD RISEN TO 8,202 FEET. AS THE DAY PROGRESSED, THEY FIRST DRIFTED SOUTH, AND THEN WEST, FALLING AND RISING NUMEROUS TIMES.

BY EIGHT IN THE EVENING THEY HAD BEGUN TO PASS OVER THE DANISH COAST AGAIN. THE BROTHERS WERE COLD, TIRED AND DEHYDRATED. EVEN SO, THEY DECIDED NOT TO LAND, BUT TO SPEND A FURTHER NIGHT IN THE AIR.

THROUGH THE SECOND NIGHT THE BALLOON GRADUALLY FLEW SOUTH, RETURNING TO GERMANY.

THE BROTHERS WERE PHYSICALLY WEAK, SHIVERING UNCONTROLLABLY, AND SUFFERING CRAMPS FROM DEHYDRATION.

BY DAWN THEY WERE HALFWAY BETWEEN HANOVER AND FRANKFURT.

WE'RE OUT OF FOOD AND WATER.

EVEN SO, THEY DECIDED TO MAKE A FINAL PUSH FOR ALTITUDE. THEIR AIM WAS TO GET ABOVE 16,404 FEET.

BUT THEY WERE SO WEAK THAT THEY WERE UNABLE TO PUSH A SACK OF BALLAST OVER THE RIM OF THE BASKET.
WE SHOULD LAND.

THE BROTHERS LANDED NEAR ASCHAFFENBURG, EAST OF FRANKFURT.

THEY HAD MANAGED TO STAY IN THE AIR FOR 52 AND A HALF HOURS, BREAKING THE WORLD RECORD BY 17 HOURS.

THIS ACHIEVEMENT WAS REPORTED IN NEWSPAPERS AROUND THE WORLD. THEY WERE FAMOUS.

BUT THIS WAS FAR FROM THE END OF ALFRED WEGENER'S ADVENTURES. LATER THAT SAME YEAR HE WAS SELECTED TO TAKE PART IN AN ARCTIC EXPEDITION.

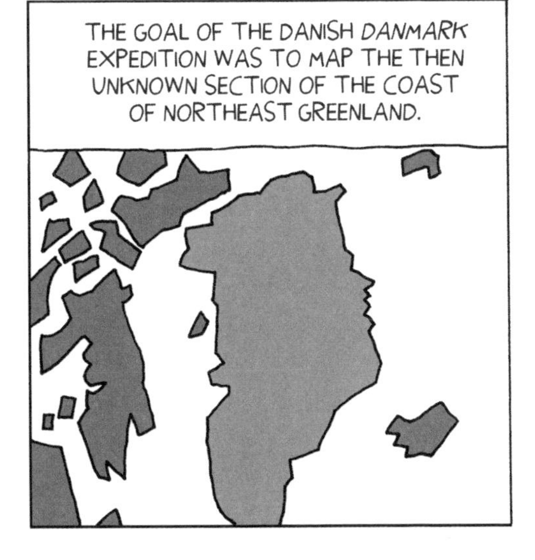
THE GOAL OF THE DANISH *DANMARK* EXPEDITION WAS TO MAP THE THEN UNKNOWN SECTION OF THE COAST OF NORTHEAST GREENLAND.

WEGENER WOULD BE AWAY FOR TWO YEARS, BUT FIRST HE HAD THE DAUNTING TASK OF TELLING HIS FATHER.
YOU'RE DOING WHAT?

RICHARD WEGENER WAS COMPLETELY OPPOSED TO THE IDEA.
YOU'RE PLANNING TO RUN OFF FOR YEARS ON AN EXPEDITION...

AT PRECISELY THE TIME WHEN YOU SHOULD BE FINDING PERMANENT PENSIONED EMPLOYMENT AND STARTING A FAMILY?

IT SOON BECAME CLEAR TO RICHARD THAT HIS SON WAS NOT GOING TO CHANGE HIS MIND.

THIS IS A LEGITIMATE SCIENTIFIC UNDERTAKING AND I'M A TRAINED SCIENTIST.

SO A SOMEWHAT BAFFLED WEGENER SENIOR OFFERED TO ADVANCE ALFRED THE MONEY HE NEEDED TO OUTFIT HIMSELF FOR THE JOURNEY.

SIGH! YOU MIGHT AS WELL MOVE BACK INTO THIS HOUSE WHILE YOU MAKE PREPARATIONS. IT WILL HELP YOU SAVE MONEY.
THANK YOU, FATHER.

JULY 1906, DANMARKSHAVN, GREENLAND. WEGENER WAS HIRED AS A METEOROLOGIST, BUT IT RAPIDLY BECAME CLEAR THAT THE EXPEDITION LEADER, LUDVIG MYLIUS-ERICHSEN, WAS NOT THAT INTERESTED IN THE KIND OF SCIENCE ALFRED WANTED TO DO.

MYLIUS-ERICHSEN'S AIM WAS TO COMPREHENSIVELY MAP THE AREA, AND TO THIS END HE AND MOST OF THE EXPEDITION MEMBERS MADE EXTENSIVE JOURNEYS ALONG THE COAST AND INTO THE INTERIOR, WHERE THERE WAS A VAST REGION OF LAKES AND FJORDS, ALL UNEXPLORED. THIS LEFT WEGENER LARGELY ALONE TO MAKE HIS METEOROLOGICAL OBSERVATIONS.

NEVERTHELESS, WHILE STRUGGLING WITH LONELINESS, DEPRESSION, EQUIPMENT FAILURE, FROSTBITE AND SHEER PHYSICAL HARD WORK, WEGENER MANAGED TO CONSTRUCT THE FIRST METEOROLOGICAL STATION IN GREENLAND.

HE LEARNED TO DRIVE A DOG-SLED TEAM AND TOOK PART IN A FEW OF THE EXPLORATORY FORAYS INLAND.

ON ONE OF THESE JOURNEYS, A 65-DAY TREK TO THE NORTHERN TIP OF THE CONTINENT, HE CAME OUT OF HIS TENT TO BE CONFRONTED BY A POLAR BEAR.

HE MANAGED TO GET HIS GUN UP IN TIME, KILLING IT WITH ONE SHOT.

THE WHOLE INCIDENT WAS COVERED IN HIS DIARY BY A SINGLE LINE.

I SAVED A GOOD PIECE OF BEAR MEAT FOR MY TEAM.

WEGENER FELT THAT HE'D PROVEN HIMSELF DURING THE *DANMARK* EXPEDITION. HE'D SURVIVED A GRUELLING AND DANGEROUS JOURNEY.

HE'D DONE GOOD SCIENCE, HUNTED MUSK OX AND BEAR, EXPLORED THE COAST AND FJORDS, AND CROSSED THE INLAND ICE.

HE'D TRANSITIONED FROM BEING THE LONE GERMAN SCIENTIST TO AN ACCEPTED MEMBER OF THE TEAM.

WEGENER'S VIRTUAL SILENCE ON THE SLEDGE TRIP WAS APPROVED OF BY THE OTHERS, WHO ADMIRED HIS UNCOMPLAINING TACITURNITY, ALTHOUGH THE REASON FOR THIS WAS LARGELY JUST HIS POOR GRASP OF DANISH.

OVERALL THE EXPEDITION TRAVELLED MORE THAN 4,000 MILES. THEY PROVED THAT THE EXISTING CHARTS WERE RADICALLY INCORRECT.

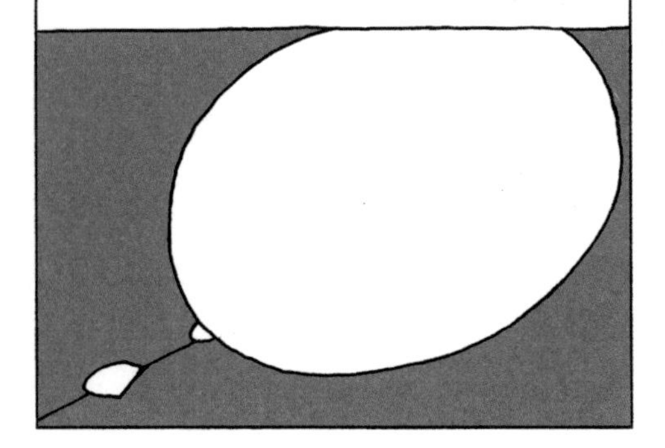

IT WAS BECAUSE THE PREVIOUS MAPS WERE WRONG THAT MYLIUS-ERICHSEN MADE A FATAL MISTAKE.

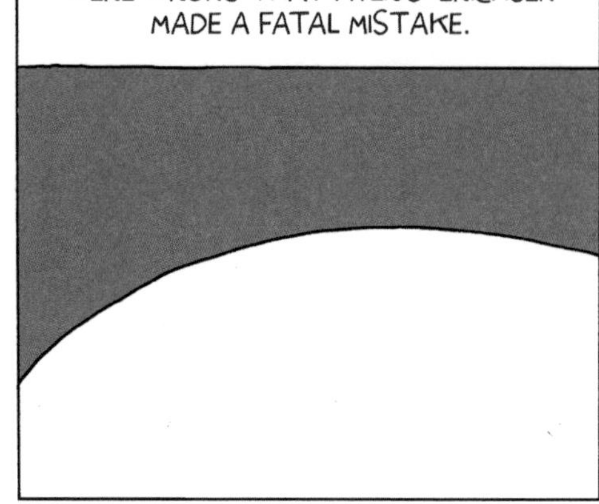

HE AND TWO OTHER MEN WERE DELAYED WHEN RETURNING FROM A TRIP NORTH.

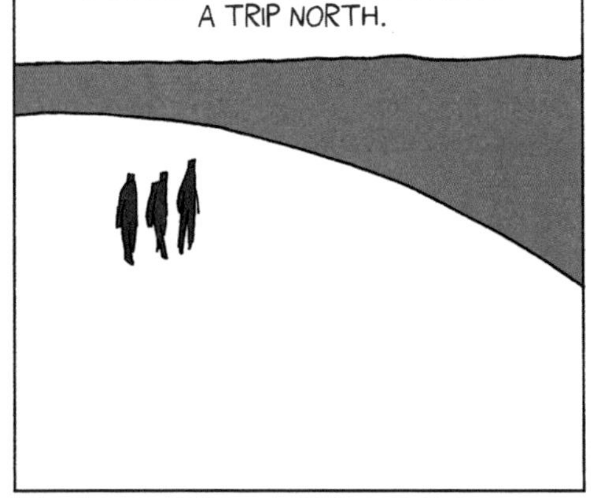

THE WINTER WAS ENDING. BY THE TIME THEY REACHED THE ICE SHEET THEY NEEDED TO CROSS IN ORDER TO REACH THE SHIP, IT WAS OPEN WATER.

THEY WERE FORCED TO SPEND THE SUMMER IN THE AREA, WITHOUT PROVISIONS, AND POORLY EQUIPPED TO HUNT FOR FOOD.

BY SEPTEMBER THE NEWLY FROZEN ICE ALLOWED THEM TO CONTINUE, BUT IT WAS TOO LATE.

ON THE JOURNEY THEY DIED OF STARVATION, EXHAUSTION AND COLD.

ON THE 21ST JULY 1908, THE *DANMARK* SAILED FOR HOME.

IN COPENHAGEN THEY WERE RECEIVED AS HEROES.

KING FREDERIK VIII PRAISED THEIR COURAGE AND DECORATED THEM ALL WITH THE DANISH ROYAL ORDER OF MERIT.

ONE OF THE FIRST PEOPLE TO VISIT THE *DANMARK* TEAM WAS THE NORWEGIAN EXPLORER ROALD AMUNDSEN.
I'M PREPARING MY OWN EXPEDITION TO THE NORTH POLE.

THIS WAS A RUSE. HE WAS NOT HEADING NORTH, BUT SOUTH, IN AN ATTEMPT TO BEAT ROBERT FALCON SCOTT TO THE SOUTH POLE.

ON HIS RETURN TO GERMANY, WEGENER TOOK A POSITION TEACHING APPLIED ASTRONOMY, METEOROLOGY AND COSMIC PHYSICS AT MARBURG UNIVERSITY.

MUCH OF THE DATA HE'D COLLECTED ON THE GREENLAND EXPEDITION WAS INCORPORATED INTO A TEXT BOOK HE WROTE ON METEOROLOGY.
THERMO-DYNAMICS OF THE ATMOSPHERE
A L WEGENER

WEGENER'S MENTOR AT THIS TIME WAS THE RUSSIAN-BORN METEOROLOGIST WLADIMIR KÖPPEN, WHO LIVED IN HAMBURG.

KÖPPEN WAS NOTABLE FOR DEVELOPING THE KÖPPEN CLIMATE CLARIFICATION SYSTEM, AS WELL AS FOR BEING CO-AUTHOR OF THE WORLD'S FIRST CLOUD ATLAS.

WEGENER SPENT MUCH OF HIS TIME AT KÖPPEN'S HOME, WHERE HE CAME TO THE ATTENTION OF THE OLDER MAN'S TEENAGE DAUGHTER, ELSE.

ALFRED'S SKIN IS SO DARK FROM EXPOSURE TO THE SUN, AND HE HAS SUCH PIERCING BLUE EYES.

DURING WINTER 1910, WHILE WEGENER WAS LOOKING THROUGH A WORLD ATLAS, IT STRUCK HIM THAT THE OUTLINES OF AFRICA AND SOUTH AMERICA LOOKED AS IF THEY COULD FIT TOGETHER LIKE THE PIECES OF A JIGSAW PUZZLE.

HE WROTE HIS THOUGHTS IN A LETTER TO ELSE.

THE FIT IS EVEN BETTER IF YOU LOOK AT A MAP OF THE FLOOR OF THE ATLANTIC AND COMPARE THE DROP-OFF INTO THE OCEAN BASIN RATHER THAN THE CURRENT EDGE OF THE CONTINENTS.

ABOUT A YEAR LATER, WEGENER HAPPENED TO BE READING A BOOK WHICH DESCRIBED FOSSIL ANIMALS THAT HAD LIVED IN WEST AFRICA AND IN BRAZIL DURING THE PALAEOZOIC ERA.

HE READ THAT SOME OF THE FOSSILS FROM BOTH REGIONS WERE ALMOST IDENTICAL.

THE BOOK'S AUTHOR EXPLAINED THIS SIMILARITY BY SAYING THAT THE CONTINENTS HAD ONCE BEEN CONNECTED BY LAND BRIDGES THAT LATER SANK BENEATH THE SEA...

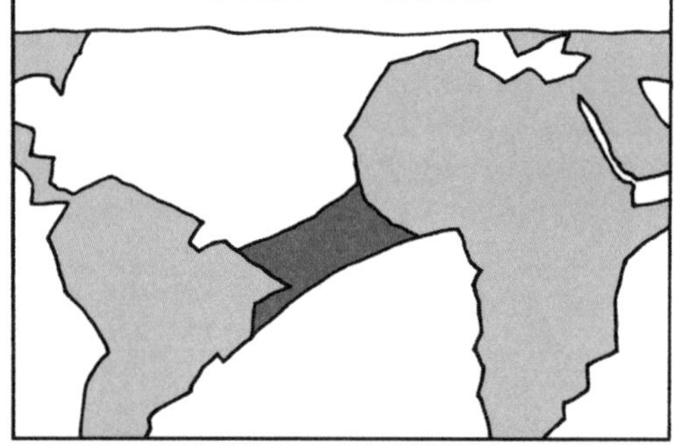

AN IDEA THAT WEGENER THOUGHT UNLIKELY. THE ALTERNATIVE NOTION HE'D PREVIOUSLY ENTERTAINED – THAT THE CONTINENTS MIGHT HAVE MOVED HORIZONTALLY ACROSS THE SURFACE OF THE GLOBE – NOW SEEMED FEASIBLE.

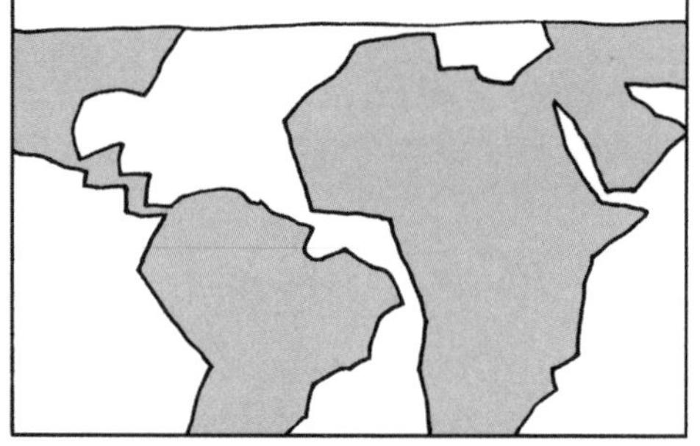

IN SLIGHTLY OVER TWO MONTHS WEGENER PRODUCED A 70-PAGE MANUSCRIPT THAT COLLATED AN ENORMOUS RANGE OF GEOLOGICAL, PALAEONTOLOGICAL AND GEOPHYSICAL EVIDENCE.

HE FIRST PRESENTED THIS NEW ARGUMENT ON THE FORMATION OF THE CONTINENTS AND OCEANS AT TWO SCIENTIFIC MEETINGS IN JANUARY 1912.

LITTLE ATTENTION WAS PAID TO WEGENER'S THEORY AT THE TIME, AND HE HIMSELF BECAME CAUGHT UP IN OTHER EVENTS.

THE FIRST OF THESE EVENTS WAS A SECOND JOURNEY TO GREENLAND, AS PART OF A FOUR-MAN TEAM LED BY JOHAN PETER KOCH, WHO'D ALSO BEEN A MEMBER OF THE *DANMARK* EXPEDITION.

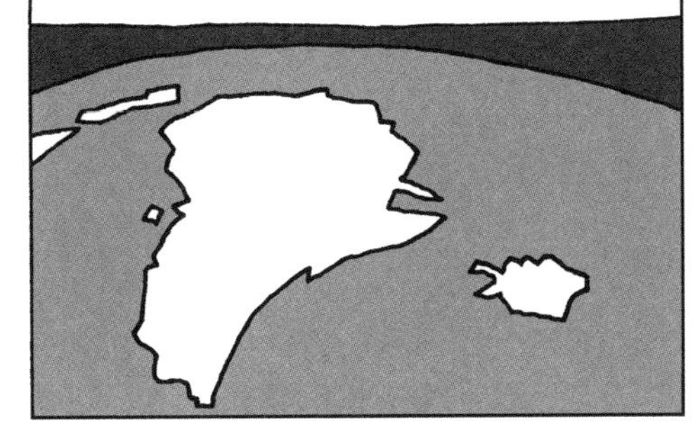

INSTEAD OF DOGS, THEY USED ICELANDIC PONIES.

WE HOPE THAT THE PONIES WILL BE ABLE TO CLIMB THE STEEP SIDES OF THE GREENLAND GLACIERS MORE EASILY THAN CAN DOG SLEDS.

BUT THE GOING WAS HARD AND DANGEROUS. THEY WERE ALMOST KILLED BY HUGE CHUNKS OF ICE FALLING FROM THE GLACIER.

KOCH FELL INTO A CREVASSE, BREAKING HIS LEG, AND WAS INCAPACITATED FOR MONTHS.
HELP!

THEY WERE THE FIRST TO WINTER ON THE INLAND ICE IN NORTHWEST GREENLAND.

AT CHRISTMAS WEGENER OPENED THE GIFT THAT ELSE HAD CAREFULLY PACKAGED FOR HIM AND FOUND TWO HAND-COLOURED PHOTOS OF HER. HE WAS DELIGHTED.

IN THE SPRING THE TEAM BEGAN THE LONG JOURNEY ACROSS THE CONTINENT.

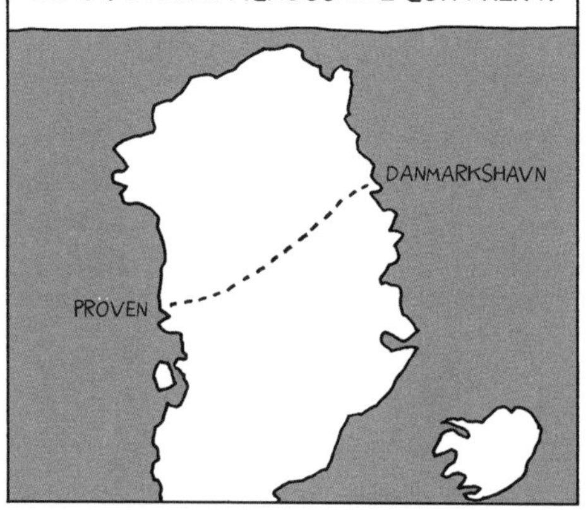


TWO MONTHS LATER, THE LAST OF THE PONIES HAVING DIED, THE EXPEDITION WAS SO STARVED THAT THEY HAD TO USE SMELLING SALTS TO KEEP THEMSELVES FROM BLACKING OUT.

SIX MILES FROM THEIR DESTINATION, PRÖVEN, THEY WERE TOO WEAK TO CONTINUE.

HAVING NOT EATEN FOR THIRTY-SIX HOURS, THEY HAD ONLY ONE CHANCE TO SURVIVE. THEY KILLED THEIR FAITHFUL DOG, GLOE.

JUST WHEN THEY WERE CUTTING UP THE HALF-COOKED DOG MEAT, WEGENER NOTICED SOMETHING OUT AT SEA.

THEIR RESCUERS WERE A DANISH PASTOR, CHEMNITZ, AND A CREW OF INUITS. CHEMNITZ KNEW THE EXPEDITION WAS DUE TO ARRIVE, SO HE'D SET OUT TO FIND THEM.

THEY WERE SAVED.

ON HIS RETURN TO GERMANY, ALFRED AND ELSE WERE MARRIED.

THEIR FIRST CHILD, HILDE, WAS BORN IN 1914.

BUT THEN...WAR.

WEGENER WAS CALLED TO ACTIVE SERVICE IN BELGIUM IMMEDIATELY...

WHILE KURT BECAME A FIGHTER PILOT.

ALFRED WAS TWICE WOUNDED – FIRST IN THE ARM AND THEN IN THE NECK. HE WAS AWARDED THE IRON CROSS (SECOND CLASS) FOR HIS COURAGE AND LEADERSHIP.

DIAGNOSED WITH A HEART CONDITION, WEGENER WAS DECLARED UNFIT TO RETURN TO THE FRONT AND WAS ASSIGNED TO THE ARMY WEATHER SERVICE...

WHERE AMONGST HIS DUTIES HE TAUGHT NAVIGATION TO ZEPPELIN CAPTAINS.

DESPITE THE WAR, HE STILL MANAGED TO WRITE THE FIRST VERSION OF HIS MOST IMPORTANT WORK – *THE ORIGIN OF CONTINENTS AND OCEANS.*

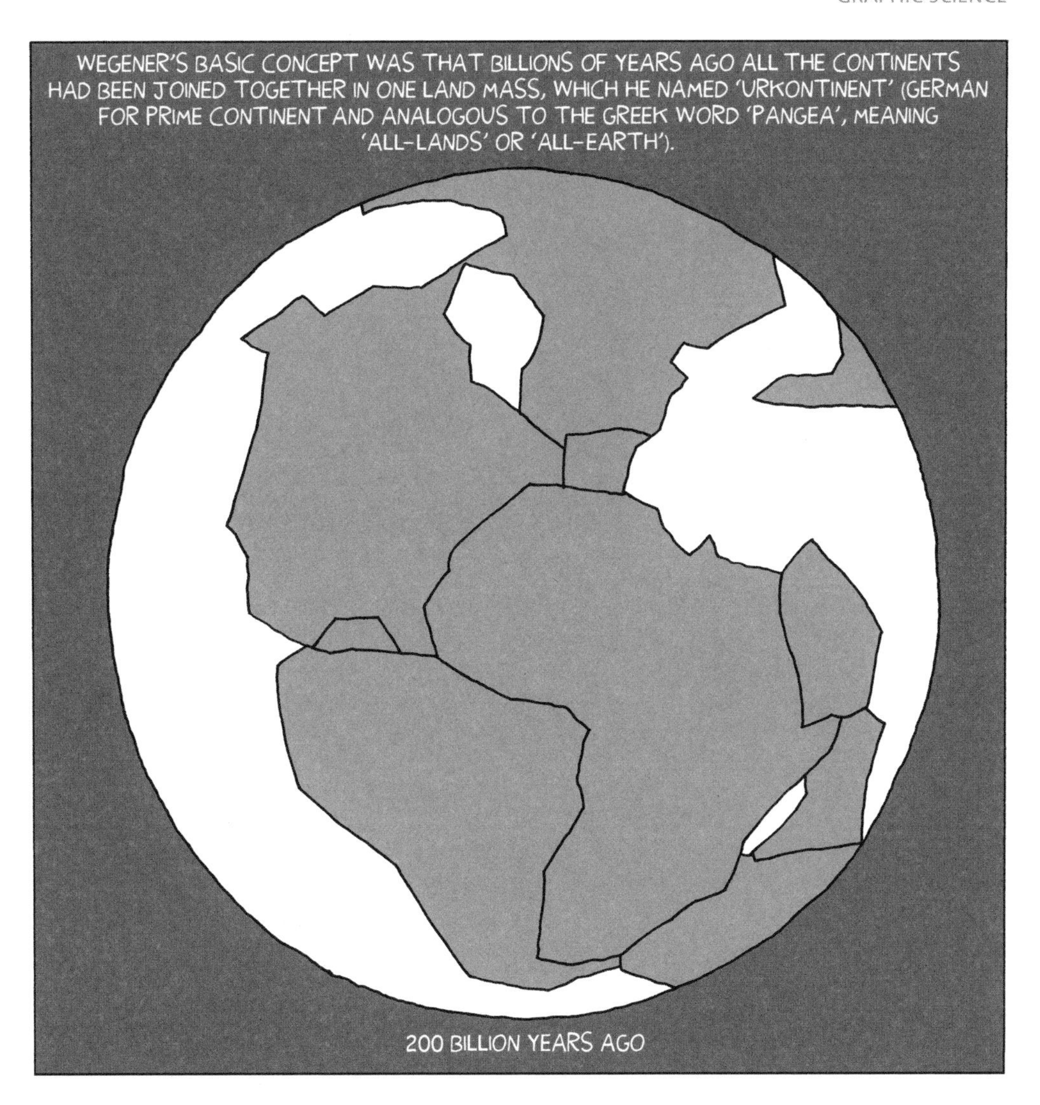
WEGENER'S BASIC CONCEPT WAS THAT BILLIONS OF YEARS AGO ALL THE CONTINENTS HAD BEEN JOINED TOGETHER IN ONE LAND MASS, WHICH HE NAMED 'URKONTINENT' (GERMAN FOR PRIME CONTINENT AND ANALOGOUS TO THE GREEK WORD 'PANGEA', MEANING 'ALL-LANDS' OR 'ALL-EARTH').
200 BILLION YEARS AGO

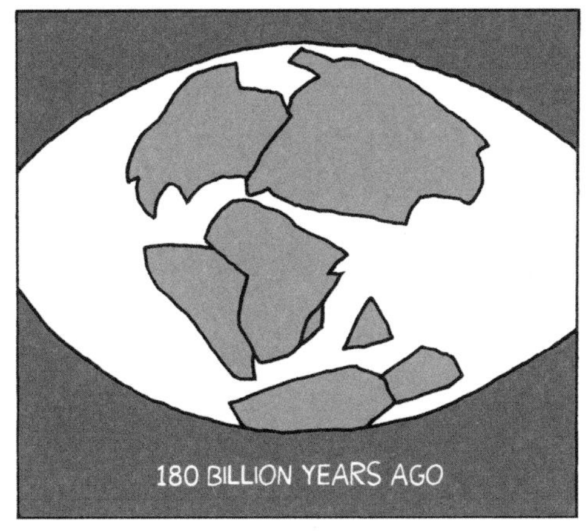
180 BILLION YEARS AGO

PRESENT DAY

AND MAKING SPACE FOR FRESH, RELATIVELY FLUID AND HOT SIMA, RISING FROM THE DEPTHS.

SIMA: THE LOWER PART OF THE EARTH'S CRUST.

SCIENTISTS WHO STRAY INTO FIELDS NOT THEIR OWN ARE OFTEN FACED WITH HOSTILITY, AND SO IT WAS WITH WEGENER.

CRITICISM OF HIS CONTINENTAL DRIFT THEORY WAS FEROCIOUS, ESPECIALLY FROM THE UNITED STATES, WHERE HIS BOOK WAS PUBLISHED AFTER THE WAR.

IF WE ARE TO BELIEVE WEGENER'S HYPOTHESIS, WE MUST FORGET EVERYTHING WHICH HAS BEEN LEARNED IN THE LAST SEVENTY YEARS, AND START ALL OVER AGAIN.

ROLLIN T. CHAMBERLIN, GEOLOGIST

MUCH OF THIS RESISTANCE CAME ABOUT BECAUSE, IF WEGENER'S THEORY WAS CORRECT...
LET ME IN.
GEOLOGY

IT WOULD DESTROY IDEAS MANY SCIENTISTS HAD SPENT THEIR ENTIRE CAREERS BELIEVING AND STUDYING.
GEOLOGY
NO!

BUT, BEYOND THESE DIFFICULTIES, EXTRAORDINARY CLAIMS REQUIRE EXTRAORDINARY EVIDENCE, AND WEGENER SIMPLY LACKED THE CONCLUSIVE PROOF HE NEEDED.
GEOLOGY
SLAM!

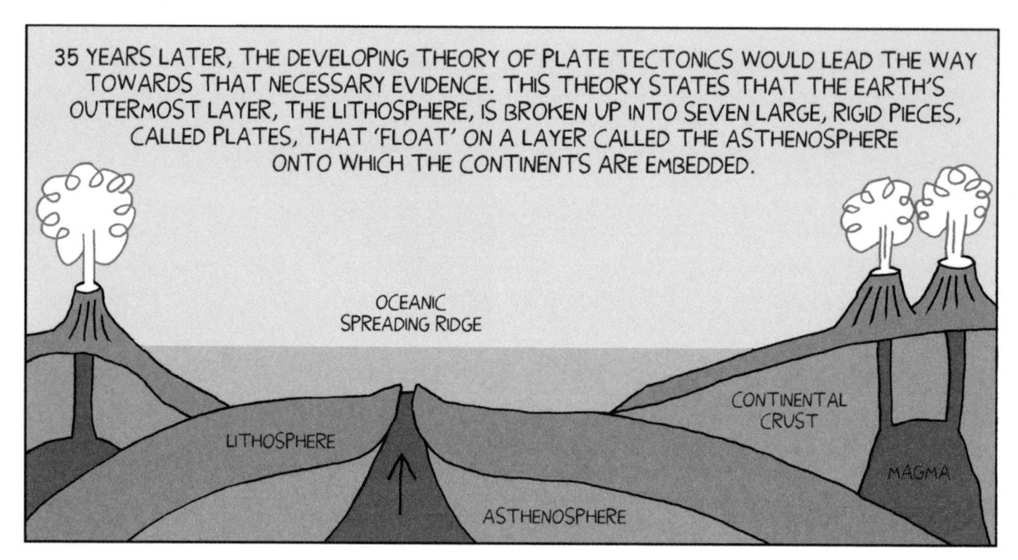
35 YEARS LATER, THE DEVELOPING THEORY OF PLATE TECTONICS WOULD LEAD THE WAY TOWARDS THAT NECESSARY EVIDENCE. THIS THEORY STATES THAT THE EARTH'S OUTERMOST LAYER, THE LITHOSPHERE, IS BROKEN UP INTO SEVEN LARGE, RIGID PIECES, CALLED PLATES, THAT 'FLOAT' ON A LAYER CALLED THE ASTHENOSPHERE ONTO WHICH THE CONTINENTS ARE EMBEDDED.
OCEANIC SPREADING RIDGE
CONTINENTAL CRUST
LITHOSPHERE
MAGMA
ASTHENOSPHERE

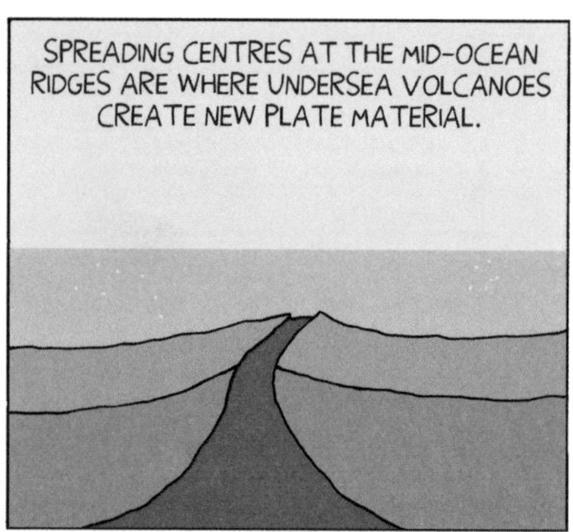
SPREADING CENTRES AT THE MID-OCEAN RIDGES ARE WHERE UNDERSEA VOLCANOES CREATE NEW PLATE MATERIAL.

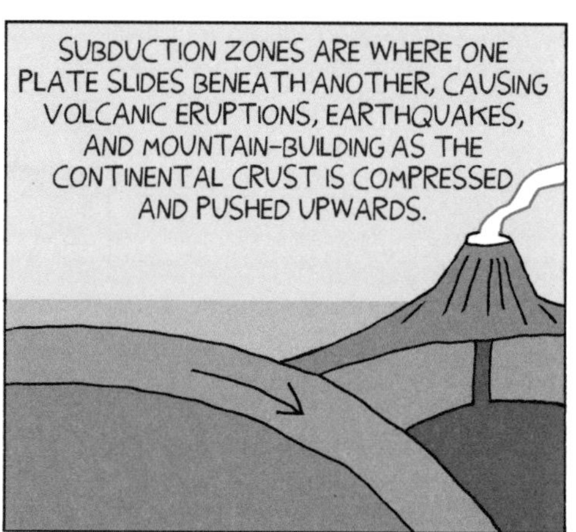
SUBDUCTION ZONES ARE WHERE ONE PLATE SLIDES BENEATH ANOTHER, CAUSING VOLCANIC ERUPTIONS, EARTHQUAKES, AND MOUNTAIN-BUILDING AS THE CONTINENTAL CRUST IS COMPRESSED AND PUSHED UPWARDS.

TWO MAIN THEORIES HAVE EMERGED AS TO WHAT CAUSES PLATE MOVEMENT.

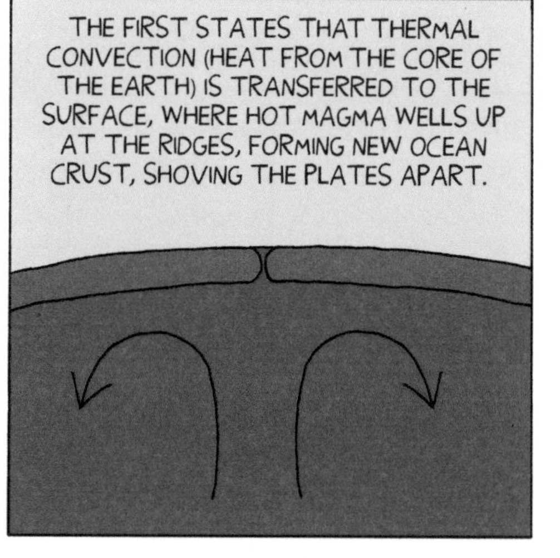
THE FIRST STATES THAT THERMAL CONVECTION (HEAT FROM THE CORE OF THE EARTH) IS TRANSFERRED TO THE SURFACE, WHERE HOT MAGMA WELLS UP AT THE RIDGES, FORMING NEW OCEAN CRUST, SHOVING THE PLATES APART.

AT SUBDUCTION ZONES, THE NOW COLD PLATE SINKS, PULLING THE CRUST ABOVE IT DOWNWARDS.

THE OTHER THEORY STATES THAT THE NEWLY FORMED PLATE AT THE MID-OCEAN RIDGE IS NOT AS DENSE AS THE ASTHENOSPHERE IT SITS ON.

THE PLATE GETS DENSER WITH AGE AS IT COOLS AND THICKENS. THIS CAUSES IT TO SINK INTO THE EARTH AT THE SUBDUCTION ZONE.

IN BOTH THEORIES THERE IS THEN FURTHER SPREADING AND RIFTING IN MID-OCEAN AS THE PLATES ARE TORN APART.

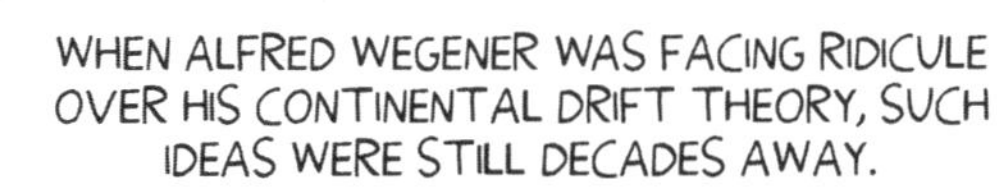
WHEN ALFRED WEGENER WAS FACING RIDICULE OVER HIS CONTINENTAL DRIFT THEORY, SUCH IDEAS WERE STILL DECADES AWAY.

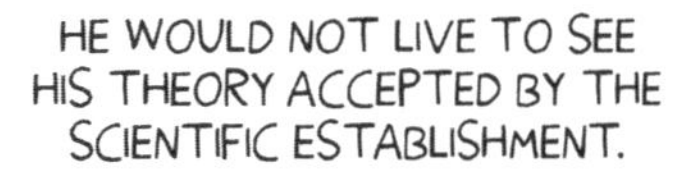
HE WOULD NOT LIVE TO SEE HIS THEORY ACCEPTED BY THE SCIENTIFIC ESTABLISHMENT.

IN 1924 WEGENER FINALLY ACHIEVED THE PROFESSORSHIP HE HAD LONG WANTED, TEACHING METEOROLOGY AND GEOPHYSICS AT THE UNIVERSITY OF GRAZ.

HOWEVER, BY 1928 HIS THOUGHTS WERE ONCE AGAIN TURNING TO GREENLAND.
HMM!
N

ELSE WAS UNIMPRESSED.
AT 48 YOU'RE SURELY TOO OLD FOR THIS SORT OF WORK?

MY DEAR, GREENLAND, IS SO WELL KNOWN NOW THAT THERE WOULD BE LITTLE POINT IN ATTEMPTING A DANGEROUS, RECORD-BREAKING EXPEDITION...

AND AS EXPEDITION LEADER ANY STRENUOUS PHYSICAL ACTIVITY WOULD FALL TO MY SUBORDINATES AND NOT TO ME.

BESIDES, DEAREST, THE TECHNOLOGY AVAILABLE TO US NOW MAKES IT FAR SAFER THAN ON THOSE EARLY ARCTIC TRIPS.
HMM!

THE AIM OF THE 1930 GREENLAND EXPEDITION WAS TO ESTABLISH THREE STATIONS FROM WHICH THE THICKNESS OF THE CONTINENT'S ICE SHEET COULD BE MEASURED AND YEAR-ROUND WEATHER OBSERVATIONS MADE.

CRUCIAL TO THE SUCCESS OF THE EXPEDITION WAS THE MID-ICE STATION, SOME 250 MILES INLAND, WHERE TWO MEN WOULD BE BASED OVER THE WINTER.

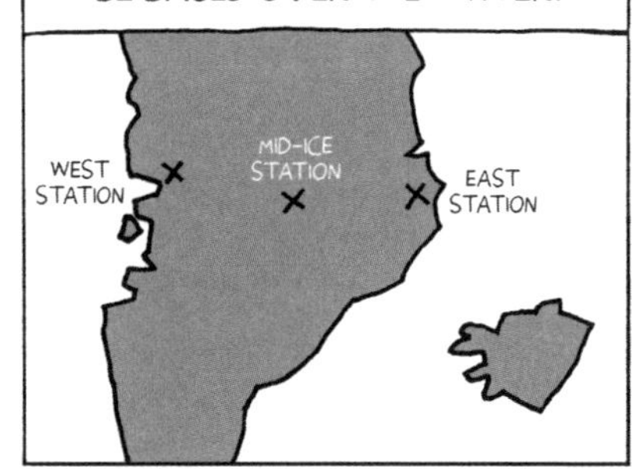


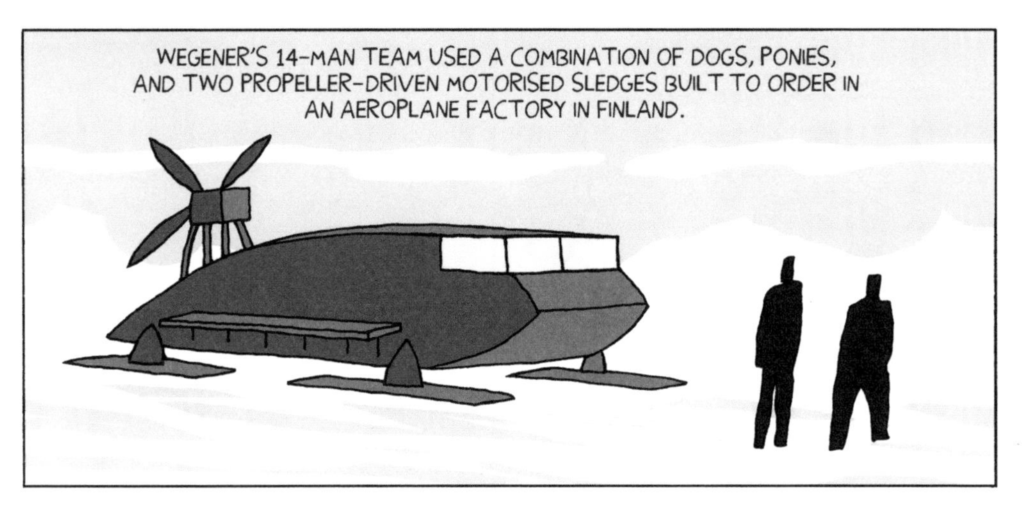

WEGENER'S 14-MAN TEAM USED A COMBINATION OF DOGS, PONIES, AND TWO PROPELLER-DRIVEN MOTORISED SLEDGES BUILT TO ORDER IN AN AEROPLANE FACTORY IN FINLAND.

THESE SLEDGES PROVED TO BE WOEFULLY UNDERPOWERED. THEY WERE UNABLE TO HAUL HEAVY LOADS EXCEPT IN THE BEST OF WEATHER.

THEY COULDN'T HANDLE STEEP SLOPES, STRONG HEADWINDS, WET SNOW, OR EXTREMELY LOW TEMPERATURES.

USELESS JUNK!

THE EXPEDITION WAS SIX WEEKS BEHIND SCHEDULE BECAUSE OF DIFFICULTIES GETTING THROUGH THE PACK ICE.

IN LATE SUMMER, JOHANNES GEORGI, ONE OF THE TWO MEN AT MID-STATION, CALCULATED THEY HAD INSUFFICIENT FOOD AND FUEL TO LAST THE WINTER.

MESSAGE FROM GEORGI.

HE REQUESTED FURTHER SUPPLIES.

HELP!

TEMPERATURES WERE DROPPING RAPIDLY WHEN WEGENER SET OUT WITH THE EXTRA PROVISIONS.

HE TOOK A TEAM OF 13 GREENLANDERS AND THE METEOROLOGIST FRITZ LOEWE WITH HIM. ALL KNEW THAT IT WAS LATE IN THE YEAR TO BE ATTEMPTING SUCH A HAZARDOUS JOURNEY.

LOEWE'S TOES BECAME SO FROST-BITTEN THAT THEY HAD TO BE AMPUTATED WITH A PENKNIFE WITHOUT ANAESTHETIC.

ARRGH!

IN THE END ONLY ONE OF THE GREENLANDERS MADE IT TO MID-STATION. THE REST TURNED BACK, THEIR REINDEER CLOTHING PROVING INADEQUATE FOR SUCH HARSH CONDITIONS.

THE REMAINING GREENLANDER WAS RASMUS VILLUMSEN. HE, ALONG WITH LOEWE AND WEGENER, REACHED THEIR DESTINATION ON 30TH OCTOBER.

THEY HAD BEEN TRAVELLING AT HIGH ALTITUDE ON THE ICE CAP AT TEMPERATURES FAR BELOW FREEZING, IN DEEP SNOW, AGAINST STRONG WINDS, FOR 40 DAYS.

THE TRAGEDY WAS THAT THE JOURNEY WAS UNNECESSARY. GEORGI HAD MISCALCULATED THE AMOUNT OF SUPPLIES AT MID-STATION.

THE STATION HAD 40 PER CENT MORE FOOD AND FUEL THAN HE REALISED. MORE THAN ENOUGH TO LAST THE WINTER.

THIS FACT, DISCOVERED LATER, WAS TO CAUSE JOHANNES GEORGI MUCH GUILT IN THE YEARS FOLLOWING.

ON 1ST NOVEMBER 1930, WEGENER CELEBRATED HIS 50TH BIRTHDAY AT MID-STATION.

THAT SAME DAY HE AND VILLUMSEN BEGAN THEIR RETURN JOURNEY TO THE COAST. LOEWE STAYED BEHIND DUE TO HIS BADLY FROSTBITTEN FEET.

GEORGI WATCHED THEM DISAPPEAR INTO THE FOG.

THE TEMPERATURE WAS BRUTALLY LOW, AROUND MINUS 50 CELSIUS (MINUS 58 FAHRENHEIT).

THEIR SITUATION DETERIORATED RAPIDLY. DOGS BEGAN TO DIE. AROUND THE HALFWAY POINT THEY FOUND THEY ONLY HAD ENOUGH DOGS LEFT TO PULL ONE SLED...

WHICH VILLUMSEN DROVE, WHILE WEGENER CONTINUED ON SKIS.

ABOUT FOUR DAYS LATER, ALFRED WEGENER DIED, MOST LIKELY FROM A HEART ATTACK. CAUSED BY THE IMMENSE PHYSICAL STRESS OF SKIING AT HIGH ALTITUDE FOR SO LONG. VILLUMSEN BURIED WEGENER IN THE SNOW AND PUSHED ON TOWARDS THE SAFETY OF THE COAST. HE WAS NEVER SEEN AGAIN.

WHEN SPRING CAME, OTHER MEMBERS OF THE EXPEDITION FOUND WEGENER'S GRAVE, WHICH VILLUMSEN HAD MARKED WITH A PAIR OF SKIS.

IN JULY, KURT WEGENER ARRIVED IN GREENLAND TO TAKE OVER LEADERSHIP OF THE EXPEDITION.

HE ARRANGED FOR A TWENTY-FOOT-HIGH IRON CROSS TO BE ERECTED OVER HIS BROTHER'S GRAVE.

BUT, IN THE YEARS SINCE, THE LOCATION OF ALFRED WEGENER'S REMAINS HAS BEEN LOST UNDER THE ICE AND SNOW.

END.

JOCELYN
BELL BURNELL

(SUSAN) JOCELYN BELL BURNELL WAS BORN INTO A QUAKER FAMILY IN BELFAST, NORTHERN IRELAND, IN 1943.

FROM THE BEGINNING, BELL BURNELL FOUND THERE WERE BARRIERS TO HER INTEREST IN SCIENCE.

IN HER FIRST WEEK AT SECONDARY SCHOOL, THE BOYS AND GIRLS WERE SEPARATED.

THE BOYS WERE TOLD TO REPORT TO THE SCIENCE LAB...

WHILE THE GIRLS WERE SENT TO THE DOMESTIC SCIENCE CLASS TO LEARN COOKING, NEEDLEWORK AND OTHER HOUSEHOLD SKILLS.
?

BELL BURNELL'S PARENTS, AS QUAKERS, WERE STRONG BELIEVERS IN THE EDUCATION OF GIRLS. THEY WERE NOT PLEASED.

THEY MADE A COMPLAINT TO THE SCHOOL, AS DID THE PARENTS OF TWO OTHER GIRLS.

AS A RESULT, THE NEXT TIME THE SCIENCE CLASS MET, THERE WERE THREE GIRLS AMONGST ALL THE BOYS.

BELL BURNELL'S FATHER WAS AN ARCHITECT WHO HAD HELPED DESIGN THE ARMAGH PLANETARIUM.

IT WAS THROUGH HER FATHER'S BOOKS THAT SHE DEVELOPED A FASCINATION WITH ASTRONOMY.
FRED HOYLE
FRONTIERS OF ASTRONOMY

IT WAS AN INTEREST THAT LED HER TO TAKE PHYSICS AT GLASGOW UNIVERSITY – THE ONLY WOMAN TO STUDY THAT SUBJECT IN HER YEAR.

HERE BELL BURNELL FOUND MORE DIFFICULTIES. EVERY TIME SHE ENTERED A CLASS OR LECTURE THEATRE, THE MALE STUDENTS WOULD STAMP THEIR FEET, CAT CALL AND WHISTLE.

SHE HAD TO LEARN NOT TO BLUSH, BECAUSE, IF SHE DID, THEY WOULD DO IT ALL THE MORE. THE EXPERIENCE OF BEING THE ONLY GIRL HAD AN ISOLATING EFFECT.

I WAS RATHER ON MY OWN THE WHOLE TIME.

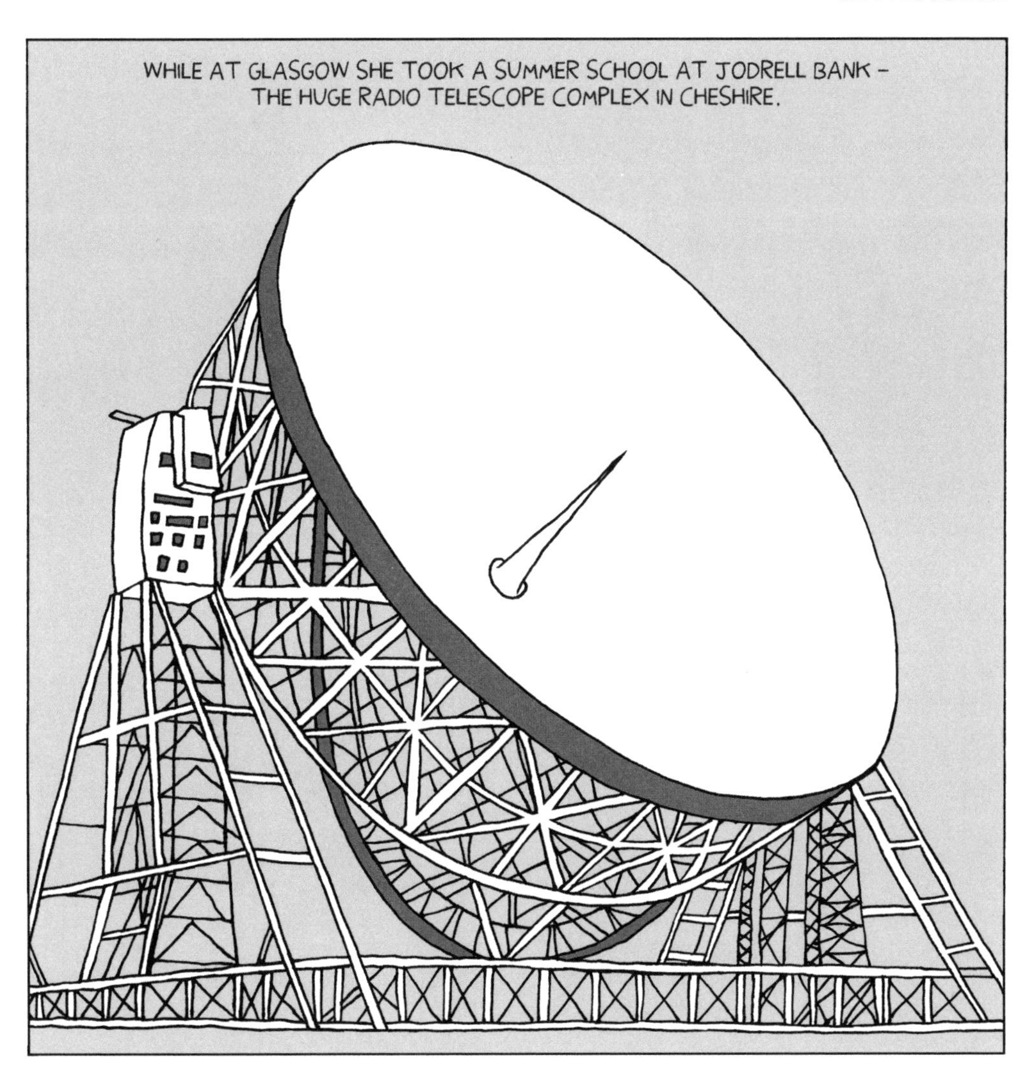
WHILE AT GLASGOW SHE TOOK A SUMMER SCHOOL AT JODRELL BANK – THE HUGE RADIO TELESCOPE COMPLEX IN CHESHIRE.

ON GRADUATION FROM GLASGOW IN 1965, BELL BURNELL HOPED TO DO HER PHD IN PHYSICS AT JODRELL BANK.

HOWEVER...
OH, THEY'LL NEVER TAKE A WOMAN HERE.

APPARENTLY THERE HAD BEEN A WOMAN STUDENT ONE TIME IN THE PAST.

SHE AND A MALE COLLEAGUE HAD PUT THE DORMITORY TO A USE IT WASN'T INTENDED.

SIR BERNARD LOVELL, THE DIRECTOR AT JODRELL BANK, GOT TO HEAR ABOUT THIS INCIDENT AND SAID...
NO WOMEN.

BELL BURNELL BELIEVED THAT SHE WASN'T ACADEMICALLY GOOD ENOUGH FOR CAMBRIDGE, WHERE STANDARDS WERE EXTREMELY HIGH.

SO SHE TURNED HER ATTENTION TO AUSTRALIA, WHERE MUCH GOOD WORK WAS BEING DONE IN RADIO ASTRONOMY.

HOWEVER, AS THE AUSTRALIAN ACADEMIC YEAR DIDN'T BEGIN UNTIL JANUARY, BELL BURNELL FOUND THAT SHE HAD A FEW MONTHS TO WAIT, SO SHE ALSO APPLIED TO CAMBRIDGE JUST IN CASE...

AND TO HER GREAT SURPRISE WAS ACCEPTED.
GOSH!

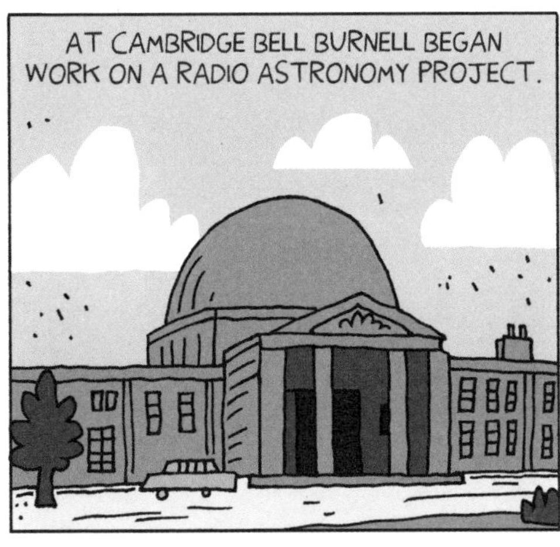
AT CAMBRIDGE BELL BURNELL BEGAN WORK ON A RADIO ASTRONOMY PROJECT.

HER PHD SUPERVISOR WAS ANTONY HEWISH.

HEWISH HAD SECURED FUNDING FOR THE CONSTRUCTION OF A LARGE RADIO TELESCOPE, OF HIS OWN DESIGN, WITH THE AIM OF STUDYING QUASARS.

SHORT FOR QUASI-STELLAR RADIO SOURCE, A QUASAR IS A MASSIVE AND EXTREMELY REMOTE CELESTIAL OBJECT.

QUASARS EMIT EXCEPTIONALLY LARGE AMOUNTS OF ENERGY. SEEN THROUGH A TELESCOPE, THEY TYPICALLY HAVE A STARLIKE APPEARANCE.

THEY ARE THE BRIGHT CENTRES OF ANCIENT GALAXIES, WHERE HUGE FORCES ARE RELEASED...

AS MATTER FALLS INTO THE BLACK HOLES THAT LIE AT THEIR HEART.

IN HER FIRST DAYS AS A GRADUATE STUDENT, BELL BURNELL WAS GIVEN A SET OF TOOLS...

COMPRISED OF WIRE-CUTTERS, PLIERS AND SCREWDRIVERS.

THE MESSAGE WAS CLEAR. ASTRONOMY WOULD NOT BE LADYLIKE WORK.

SHE SPENT THE FIRST TWO YEARS OF HER PHD IN THE FIELD, AND ONE FIELD IN PARTICULAR.

NOT ALL RADIO TELESCOPES LOOK LIKE GIANT DISHES. IN AN AREA EQUIVALENT TO TWO FOOTBALL PITCHES, BELL BURNELL AND AROUND HALF A DOZEN OTHERS BUILT WHAT WAS KNOWN AS THE INTERPLANETARY SCINTILLATION ARRAY.

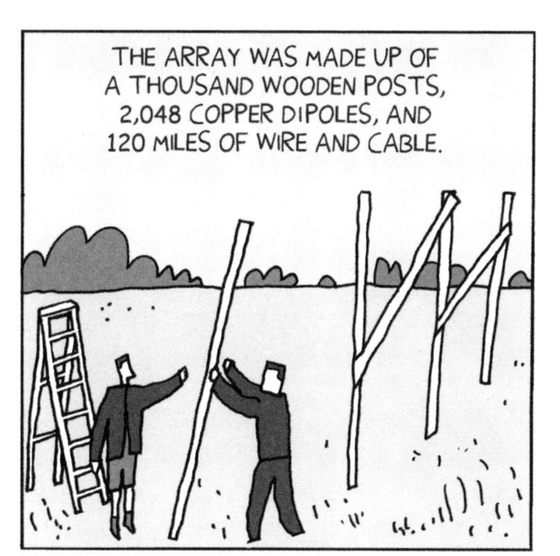
THE ARRAY WAS MADE UP OF A THOUSAND WOODEN POSTS, 2,048 COPPER DIPOLES, AND 120 MILES OF WIRE AND CABLE.

ALTHOUGH BELL BURNELL LEARNED HOW TO SWING A SLEDGEHAMMER, HER PRIME RESPONSIBILITY WAS CABLES AND PLUGS.

JUST AS STARLIGHT TWINKLES WHEN IT PASSES THROUGH THE ATMOSPHERE, A SIMILAR EFFECT HAPPENS TO RADIO WAVES WHEN THEY PASS THROUGH THE SOLAR WIND (THE STREAM OF CHARGED PARTICLES EJECTED BY THE SUN).

RADIATION FROM A COMPACT SOURCE, SUCH AS A QUASAR, WOULD SCINTILLATE, OR TWINKLE, MORE THAN A DIFFUSED EXTENDED OBJECT LIKE A GALAXY.

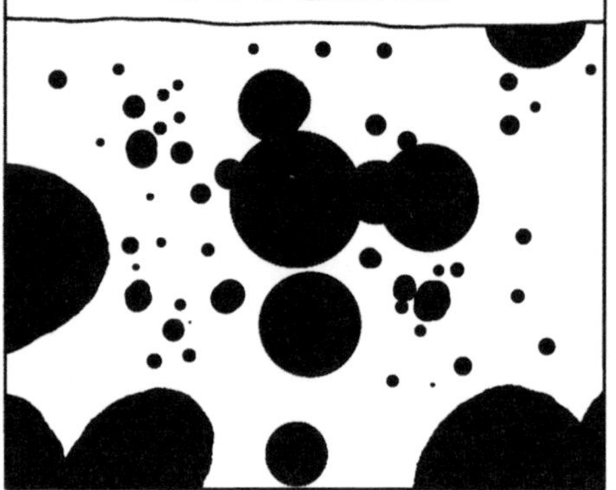
SO, BY SCANNING THE SKY FOR SCINTILLATING OBJECTS, ANTONY HEWISH HOPED TO PICK OUT LOTS OF QUASARS.

AFTER TWO YEARS, WHEN THE TELESCOPE WAS FINALLY COMPLETE, THE REST OF THE CONSTRUCTION TEAM MOVED ON TO OTHER PROJECTS...

LEAVING BELL BURNELL, THE RESEARCH STUDENT, TO OPERATE THE TELESCOPE.

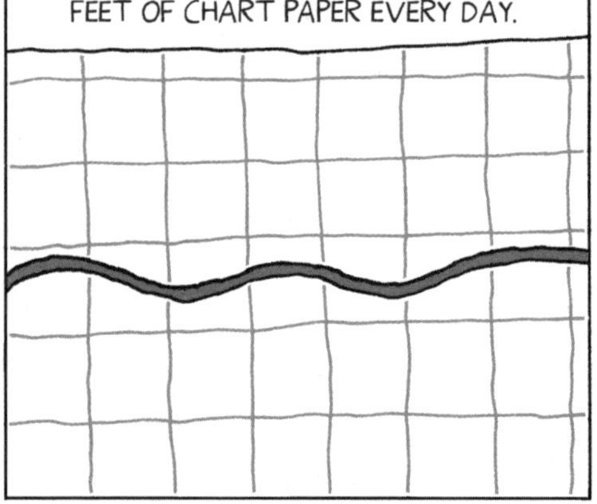
OUTPUT WAS PRINTED WITH A RED PEN ON MOVING PAPER. THE ARRAY PRODUCED 100 FEET OF CHART PAPER EVERY DAY.

ONE COMPLETE SCAN OF THE SKY TOOK FOUR DAYS AND 400 FEET OF PAPER. BELL BURNELL RAN THE TELESCOPE FOR SIX MONTHS, WHICH GAVE HER OVER THREE MILES OF PAPER TO ANALYSE.

I WOULD GO THROUGH THE CHART AND I WOULD LOG WHERE I THOUGHT I SAW QUASARS, AND I ALSO SAW THERE WERE CHUNKS OF MAN-MADE INTERFERENCE.

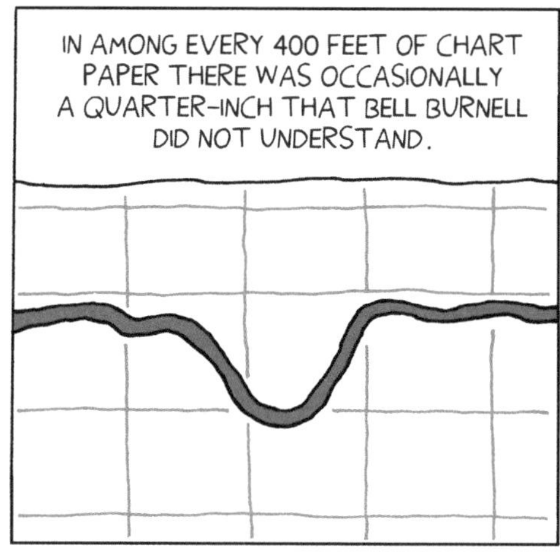
IN AMONG EVERY 400 FEET OF CHART PAPER THERE WAS OCCASIONALLY A QUARTER-INCH THAT BELL BURNELL DID NOT UNDERSTAND.

THIS BIT OF 'SCRUFF', AS BELL BURNELL CALLED IT, DIDN'T LOOK LIKE EITHER A QUASAR OR MAN-MADE INTERFERENCE.

AFTER SHE'D SEEN IT A FEW TIMES FROM THE SAME PART OF THE SKY, SHE DREW IT TO THE ATTENTION OF HEWISH.

AS THE SIGNAL WAS CRAMMED INTO A QUARTER-INCH SECTION OF THE CHART, THEY NEEDED TO SPREAD IT OUT SO IT COULD BE SEEN IN MORE DETAIL.

THE ONLY WAY TO DO THIS WAS TO RUN THE PAPER FASTER UNDER THE PEN.

IT WASN'T POSSIBLE TO RUN THE CHART PAPER AT SPEED FOR 24 HOURS. THERE WOULDN'T BE ENOUGH PAPER FOR THAT.

SO BELL BURNELL, BEING THE RESEARCH STUDENT, HAD TO GO TO THE OBSERVATORY EACH DAY AT THE APPROPRIATE TIME...

AND SWITCH THE MACHINE TO HIGH-SPEED FOR THE DURATION OF THE TRANSIT.

FOR A MONTH SHE DID THIS...

AND GOT NOTHING.

AS THE MONTH OF NON-RESULTS PROCEEDED, HEWISH BECAME IRRITATED.
IT'S A FLARE STAR AND IT'S BEEN AND GONE AND DONE IT AND YOU'VE MISSED IT.

FINALLY THE SIGNAL APPEARED AGAIN.
IT'S A SERIES OF PULSES, ONE-AND-A-THIRD SECONDS APART.

WELL, THAT SETTLES IT. IT MUST BE MAN-MADE.

HEWISH, THE EXPERIENCED ASTROPHYSICIST, REALISED THAT 1.3 SECONDS WAS TOO SMALL FOR A STAR. IT HAD TO BE INTERFERENCE.

THIS WAS AN ANXIOUS TIME FOR BELL BURNELL.

I HAD WIRED UP THIS RADIO TELESCOPE AND WAS SCARED THAT I HAD LITERALLY GOT SOME WIRES CROSSED...

THAT MY STUPIDITY WAS ABOUT TO BE DISCOVERED BY THE COMBINED BRAINS OF CAMBRIDGE, AND I MIGHT BE LEAVING WITHOUT A PHD.

ULTIMATELY THE RADIO TELESCOPE AND RECEIVER WERE EXONERATED WHEN ANOTHER TELESCOPE OPERATING ON THE SAME FREQUENCY, ALSO PICKED UP THE SIGNAL.

IT WAS A DISCUSSION BETWEEN TONY HEWISH, MARTIN RYLE (THE HEAD OF THE GROUP), AND PROBABLY JOHN SHAKESHAFT (ONE OF THE OTHER SENIOR MEMBERS OF THE GROUP).

THEY WERE ATTEMPTING TO DECIDE HOW THEY WERE GOING TO PUBLISH THE RESULTS.

IT HAD BECOME COMMON TO REFER TO THE SIGNAL AS 'LITTLE GREEN MEN', ALTHOUGH NO ONE SERIOUSLY THOUGHT IT ORIGINATED WITH ALIEN LIFE.

IT APPEARED TO BE COMING FROM AN AREA ABOUT 200 LIGHT YEARS AWAY – FAR BEYOND OUR SOLAR SYSTEM, BUT WITHIN OUR GALAXY.

ITS RAPID PULSING SUGGESTED THAT IT WAS MUCH SMALLER THAN A STAR...

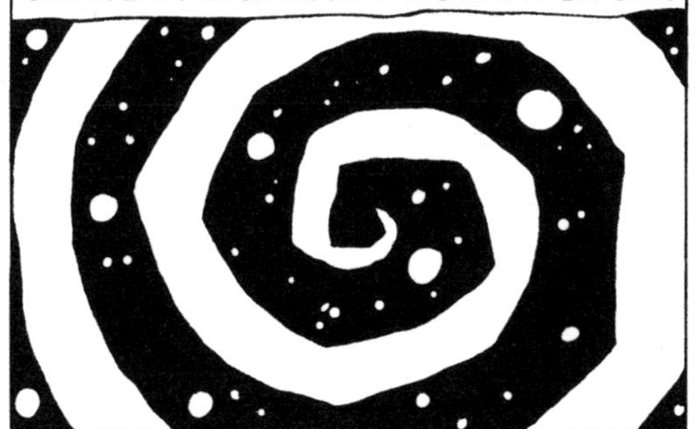
BUT THE ACCURACY OF THAT PULSING NEVER VARIED, WHICH INDICATED THAT THE OBJECT MUST HAVE HUGE RESERVES OF ENERGY, AS IF IT WERE VERY LARGE. IT WAS A PUZZLE. NO ONE HAD EVER SEEN ANYTHING LIKE IT BEFORE.

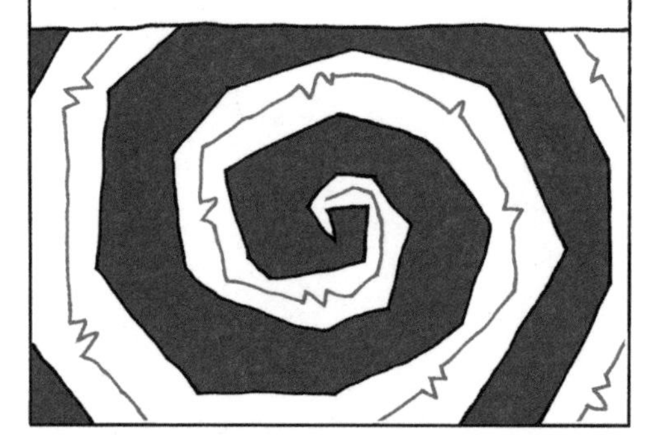
WHILE ALL THIS HAD BEEN HAPPENING, A BACKLOG OF 2,500 FEET OF CHART HAD BUILT UP THAT NEEDED ANALYSING.

IT WAS WHEN CHECKING THESE RESULTS THAT BELL BURNELL SAW SOMETHING FAMILIAR.

ANOTHER SCRUFFY BIT OF SIGNAL...

COMING FROM A COMPLETELY DIFFERENT PART OF THE SKY.

THIS FINDING RULED OUT ANYTHING LOCAL AND MADE ALIEN LIFE AS A CAUSE EVEN MORE DOUBTFUL.

IT WAS HIGHLY UNLIKELY THAT THERE WOULD BE TWO LOTS OF LITTLE GREEN MEN ON OPPOSITE SIDES OF THE UNIVERSE, BOTH DECIDING TO SIGNAL TO A RATHER INCONSPICUOUS STAR ON A RATHER CURIOUS FREQUENCY...

AND USING A TECHNIQUE THAT WAS NOT AT ALL INTELLIGENT. IT HAD TO BE SOME NEW KIND OF STELLAR SOMETHING, AND WE'D FOUND THE FIRST ONE.

OVER CHRISTMAS, HEWISH HAD KEPT THE SURVEY RUNNING, PUTTING FRESH PAPER IN THE CHART RECORDER, AND INK IN THE INKWELL.

HE THEN PILED THE UNANALYSED CHARTS ON BELL BURNELL'S DESK.

ON HER RETURN, BELL BURNELL FOUND TWO MORE OF THE MYSTERIOUS SIGNALS IN THE MILES OF CHARTS.

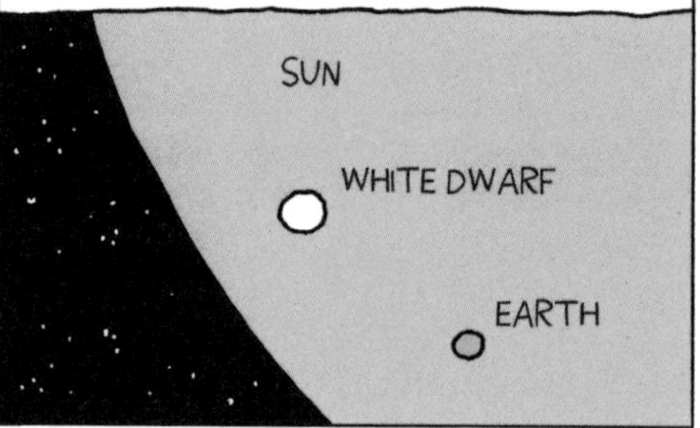
NO ONE YET KNEW WHAT THESE OBJECTS WERE. ONE THEORY PROPOSED THAT THEY MIGHT BE COMPACT STARS, KNOWN AS WHITE DWARFS – OBJECTS WITH THE MASS OF OUR SUN THAT HAVE COLLAPSED TO THE SIZE OF THE EARTH.
SUN
WHITE DWARF
EARTH

AT THE END OF JANUARY 1968, THE PAPER ANNOUNCING THE CAMBRIDGE GROUP'S DISCOVERY WAS SUBMITTED TO THE SCIENCE JOURNAL *NATURE*.

A FEW DAYS BEFORE THE PAPER WAS SUBMITTED, ANTONY HEWISH GAVE A SEMINAR IN CAMBRIDGE ABOUT THE GROUP'S FINDINGS.

AMONG THOSE ATTENDING THE SEMINAR WAS THE NOTED ASTRONOMER FRED HOYLE.

THIS IS THE FIRST I'VE HEARD OF THESE OBJECTS. I DON'T THINK IT'S A WHITE DWARF. I THINK IT'S A SUPERNOVA REMNANT.

IT SHOWS THE CALIBRE OF THE MAN AND HIS ASTROPHYSICAL UNDERSTANDING THAT, WHEN PRESENTED WITH SOMETHING LIKE THAT, WITHIN THE HOUR HE COULD HIT THE RIGHT CONCLUSION.

IN THE PAPER WE HAD BEEN A BIT AMBIGUOUS ABOUT WHAT WE HAD WITNESSED, BECAUSE WE HONESTLY DIDN'T KNOW WHAT IT WAS.

THERE WAS A LOT OF PUBLICITY FOLLOWING THE ANNOUNCEMENT. THE CAMBRIDGE TEAM FOUND THEMSELVES SUDDENLY IN THE SPOTLIGHT.

TYPICALLY THE PRESS WOULD ASK HEWISH ABOUT THE ASTROPHYSICAL IMPORTANCE OF THE DISCOVERY...

WHILE BELL BURNELL WAS ASKED...
WHAT ARE YOUR VITAL STATISTICS? DO YOU HAVE MANY BOYFRIENDS?

IT WAS THE SCIENCE CORRESPONDENT OF THE *DAILY TELEGRAPH*, ANTHONY MICHAELIS, WHO NAMED THESE MYSTERIOUS OBJECTS.

WELL, THERE ARE *QUASARS*. SO WHY DON'T YOU CALL THEM PULSARS FOR PULSATING STELLAR RADIO OBJECTS?

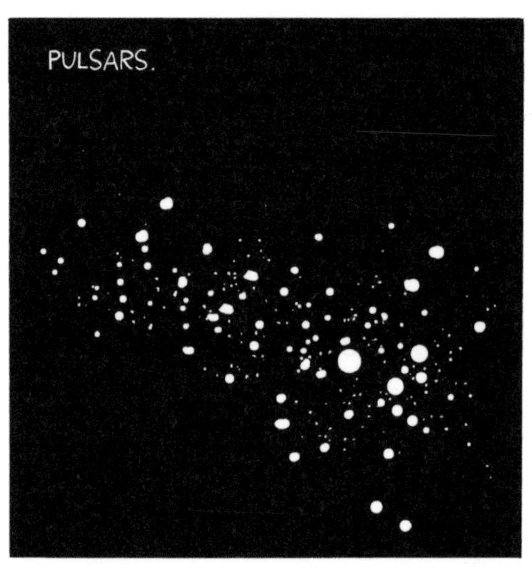
PULSARS.

WHAT ARE THEY?

WHEN A STAR WITH FOUR TO EIGHT TIMES THE MASS OF OUR SUN DIES...

IT DETONATES AS A SUPERNOVA.

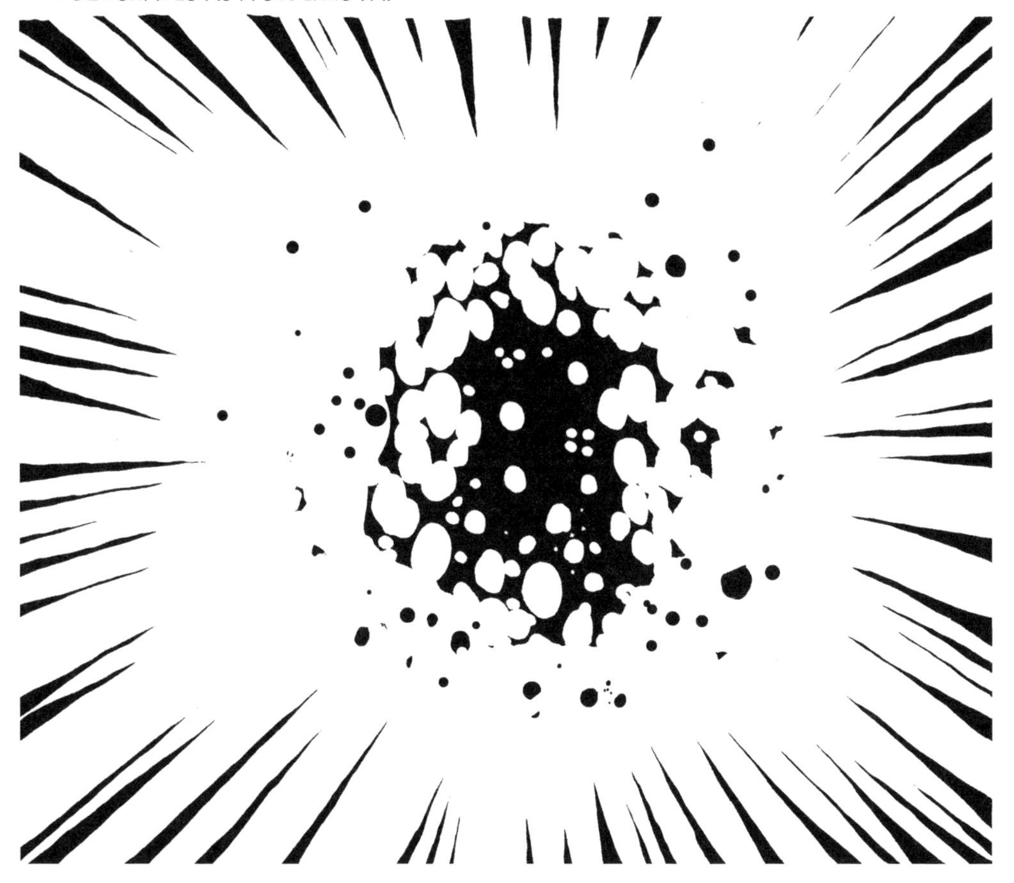

THE STAR'S OUTER LAYERS ARE BLASTED OFF INTO SPACE...

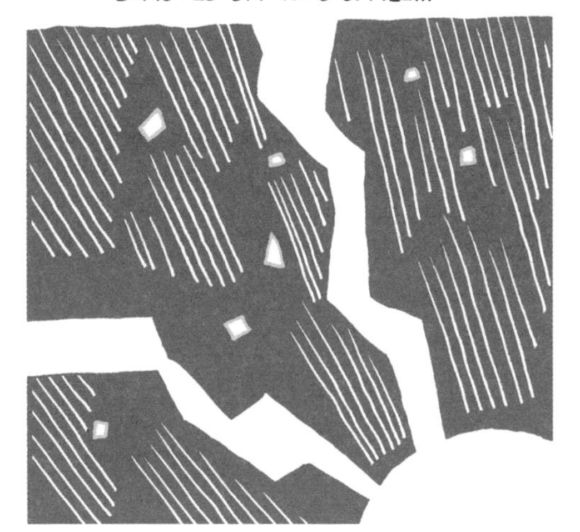

WHILE THE INNER CORE CONTRACTS DOWN WITH GRAVITY.

THE GRAVITATIONAL PRESSURE IS SO STRONG THAT IT OVERCOMES THE BONDS THAT KEEP ATOMS APART.

ELECTRONS AND PROTONS ARE CRUSHED TOGETHER TO FORM NEUTRONS.

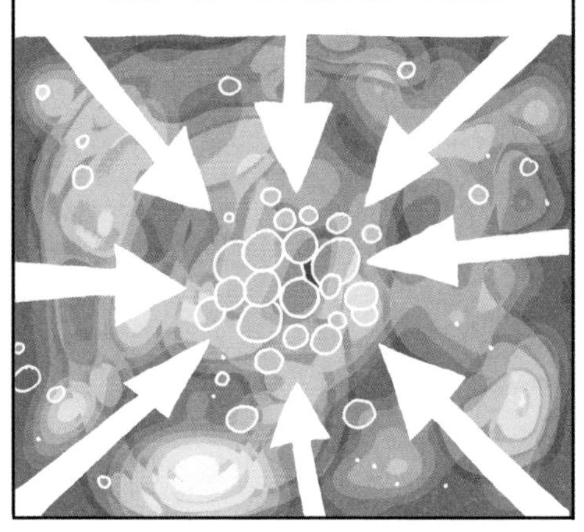

IMAGINE A STAR THE SIZE OF OUR SUN COMPRESSED INTO A SPHERE ONLY TEN KILOMETRES IN RADIUS.

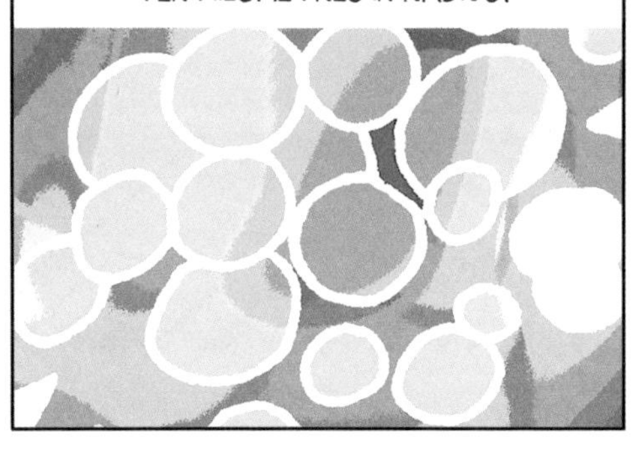

THE GRAVITY ON THIS NEUTRON STAR IS STRONG ENOUGH TO BEND LIGHT.

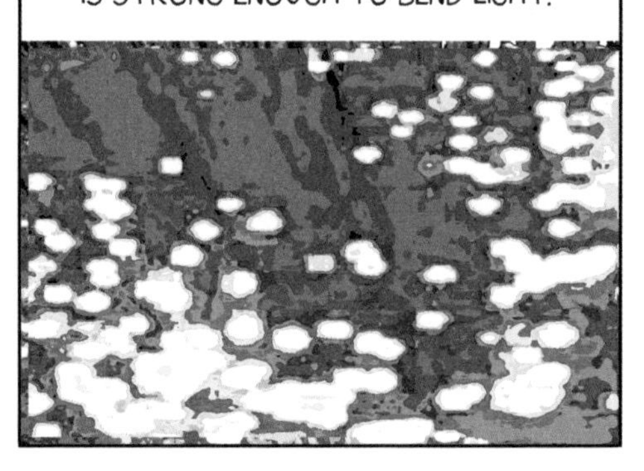

TO CLIMB HALF AN INCH IS THE EQUIVALENT OF CLIMBING MOUNT EVEREST.

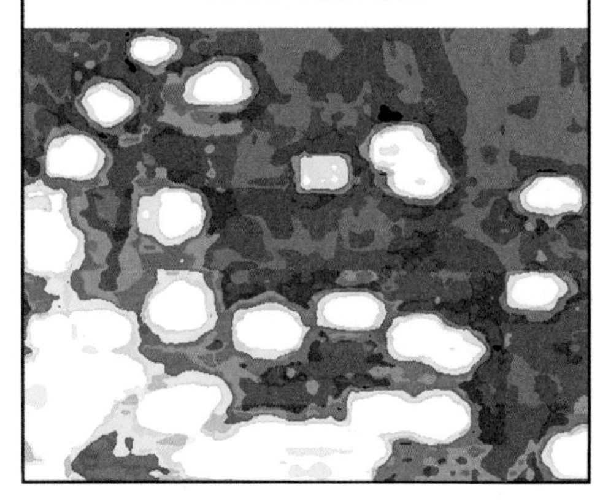

WHAT SETS PULSARS APART FROM REGULAR NEUTRON STARS IS THAT THEY'RE HIGHLY MAGNETISED AND ROTATE AT ENORMOUS SPEED.

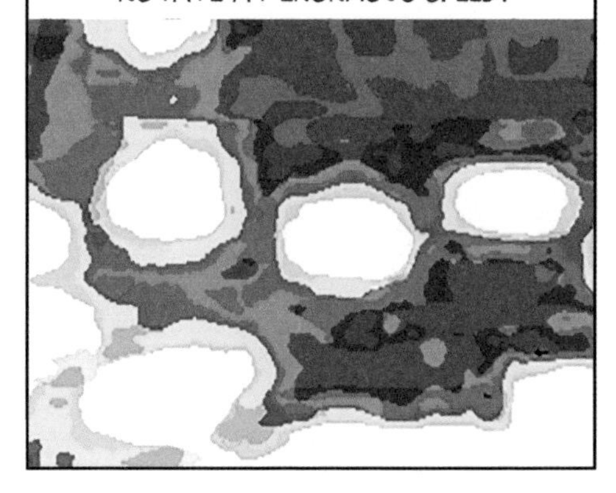

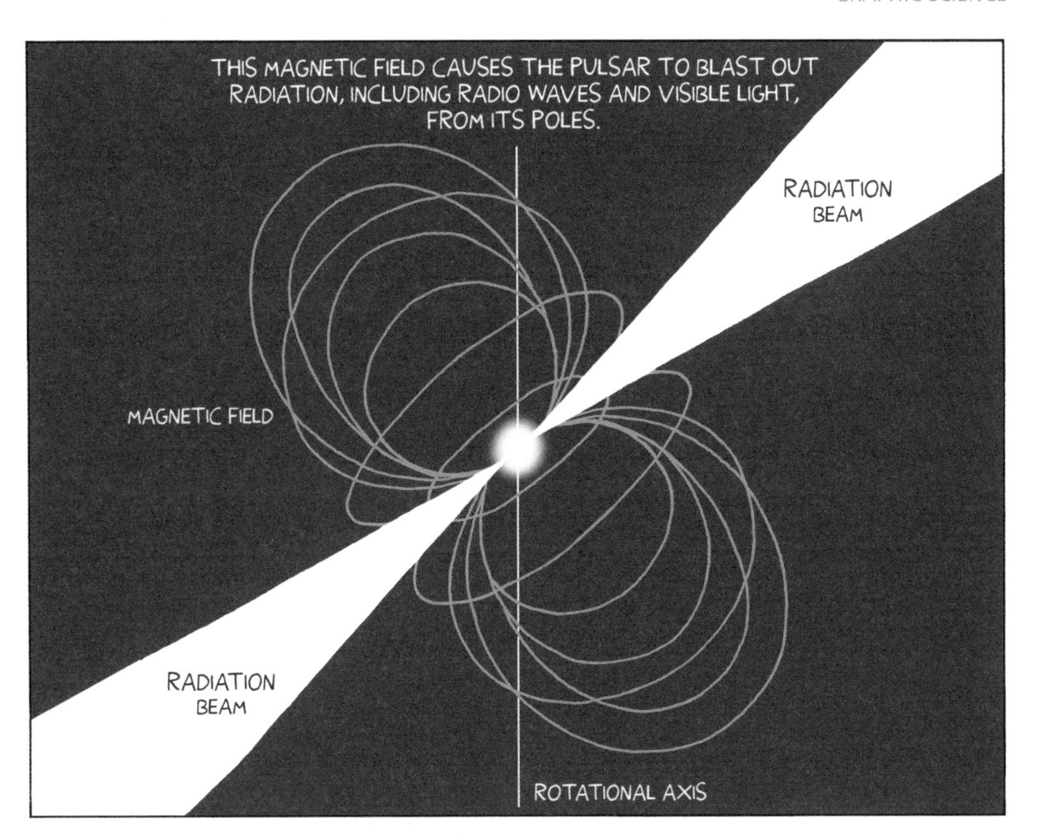
THIS MAGNETIC FIELD CAUSES THE PULSAR TO BLAST OUT RADIATION, INCLUDING RADIO WAVES AND VISIBLE LIGHT, FROM ITS POLES.
RADIATION BEAM
MAGNETIC FIELD
RADIATION BEAM
ROTATIONAL AXIS

IF THE PULSAR IS ALIGNED SO THAT THE POLES FACE THE EARTH...

THEN WE WILL SEE A FLASH OF RADIATION EVERY TIME ONE OF THESE POLES ROTATES INTO OUR FIELD OF VISION.

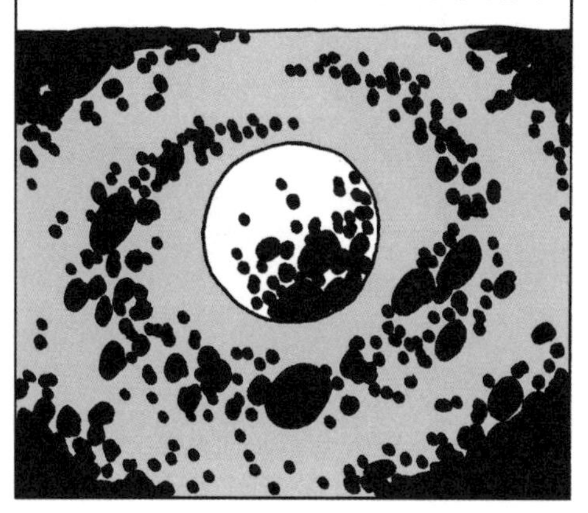

PULSARS CAN SPIN AT SPEEDS THAT ARE HARD FOR THE HUMAN MIND TO GRASP.

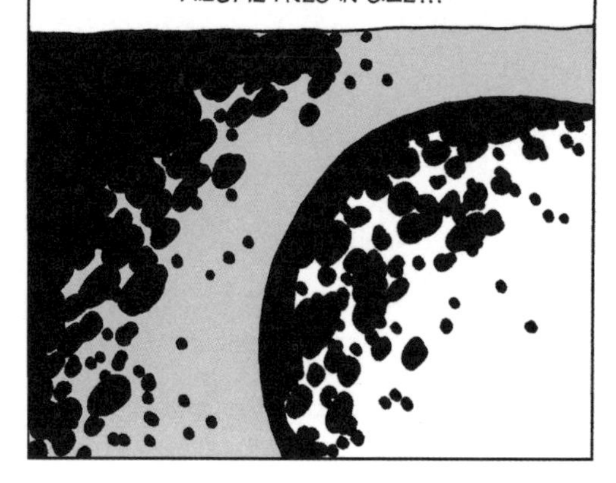

THE PULSAR PSRJ1748-2446AD, WHICH IS THOUGHT TO BE ONLY SIXTEEN KILOMETRES IN SIZE...

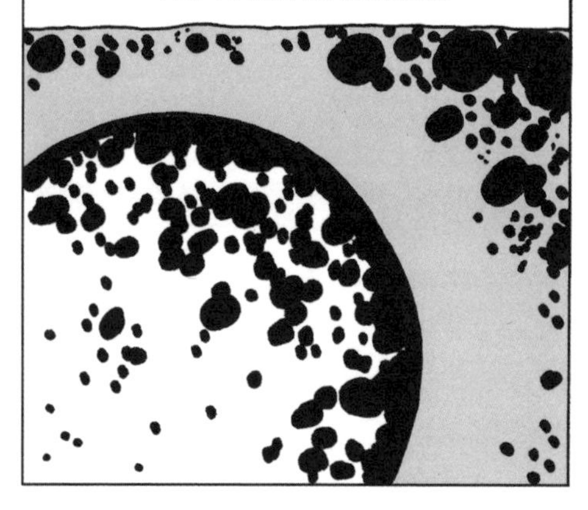

ROTATES AT AN INCREDIBLE 716 TIMES PER SECOND.

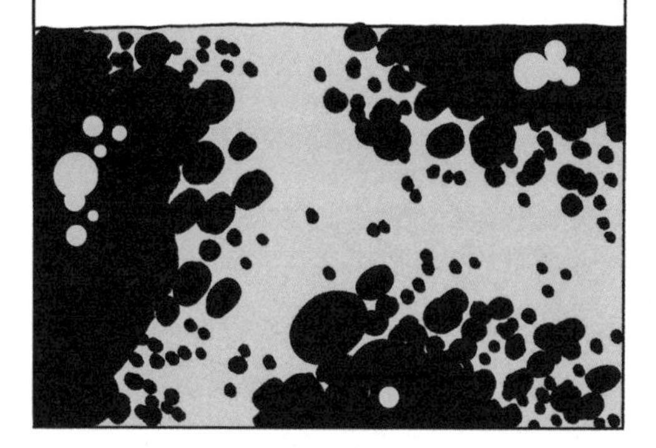

SINCE BELL BURNELL DISCOVERED THE FIRST FEW PULSARS, OVER A THOUSAND MORE HAVE BEEN FOUND.

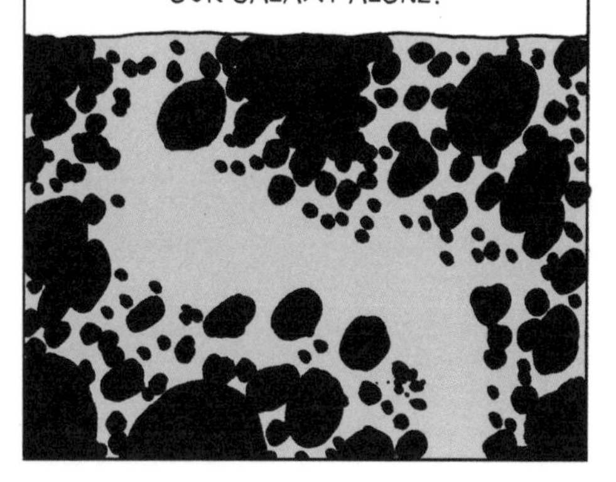

OVERALL IT IS ESTIMATED THAT THERE ARE AROUND 200,000 PULSARS IN OUR GALAXY ALONE.

IN 1968, WHEN THE PAPER ANNOUNCING THE DISCOVERY OF PULSARS WAS PUBLISHED, FIVE NAMES WERE LISTED AS CO-AUTHORS.

Observation of a Rapidly Pulsat

by

A. HEWISH
S. J. BELL
J. D. H. PILKINGTON
P. F. SCOTT
R. A. COLLINS

Mullard Radio Astronomy Observatory,
Cavendish Laboratory,
University of Cambridge

Unusual signals from pulsati
the Mullard Radio Astronom
come from local objects w
with oscillations of white d

In July 1967, a large radio telescope operating at a frequency of 81·5 MHz was brought into use at the Mullard Radio Astronomy Observatory. This instrument was designed to investigate the angular structure of compact radio sources by observing the scintillation caused by the irregular structure of the

BELL BURNELL'S NAME WAS SECOND ON THE PAPER.

by

A. HEWISH
S. J. BELL
J. D. H. PILKINGTON
P. F. SCOTT
R. A. COLLINS

Mullard Radio Astronomy Observatory,
Cavendish Laboratory,
University of Cambridge

In July 1967, a large radio telescop

SOME MAY HAVE BEEN SURPRSED WHEN IN 1974 THE NOBEL PRIZE FOR PHYSICS WAS AWARDED TO ANTONY HEWISH AND MARTIN RYLE...

WITHOUT BELL BURNELL AS CO-RECIPIENT.

IN THEIR PRESS RELEASE, THE ROYAL SWEDISH ACADEMY OF SCIENCE CITED HEWISH AND RYLE FOR THEIR PIONEERING WORK IN RADIO ASTRONOMY...

WITH PARTICULAR MENTION OF RYLE'S WORK ON APERTURE-SYNTHESIS...

AND HEWISH'S DECISIVE ROLE IN THE DISCOVERY OF PULSARS.

AT THAT TIME, SCIENCE WAS PERCEIVED AS BEING DONE BY MEN, SENIOR MEN...

MAYBE WITH A WHOLE FLEET OF MINIONS UNDER HIM WHO DID HIS BIDDING AND WEREN'T EXPECTED TO THINK.

I BELIEVE THE NOBEL PRIZE COMMITTEE DIDN'T EVEN KNOW I EXISTED.

IN A BBC INTERVIEW IN 2010, ANTONY HEWISH OUTLINED HIS THOUGHTS ON THE SUBJECT.

MY ANALOGY REALLY IS A LITTLE BIT LIKE WHEN YOU PLAN A SHIP OF DISCOVERY, AND YOU GO OFF AND SOMEBODY ON THE MASTHEAD SAYS, 'LAND HO!'

THAT'S GREAT, BUT I MEAN, WHO ACTUALLY INSPIRED IT AND DECIDED WHAT TO DO, AND SO ON? THERE IS A DIFFERENCE BETWEEN SKIPPER AND CREW.

TO BE HONEST, I DON'T THINK IT WOULD HAVE MATTERED WHO'D BEEN MY STUDENT. IT WAS A SERENDIPITOUS DISCOVERY BECAUSE SUCH A PIECE OF EQUIPMENT HAD BEEN SET UP.

I MEAN, THE DISCOVERY OF PULSARS WAS UNAVOIDABLE ONCE THE SURVEY HAD BEGUN.

CRITICS OF THE NOBEL PRIZE COMMITTEE'S DECISION HAD A DIFFERENT OPINION.

IT WOULD HAVE BEEN VERY EASY FOR BELL BURNELL TO HAVE IGNORED THE 'BITS OF SCRUFF', THEY SAID...

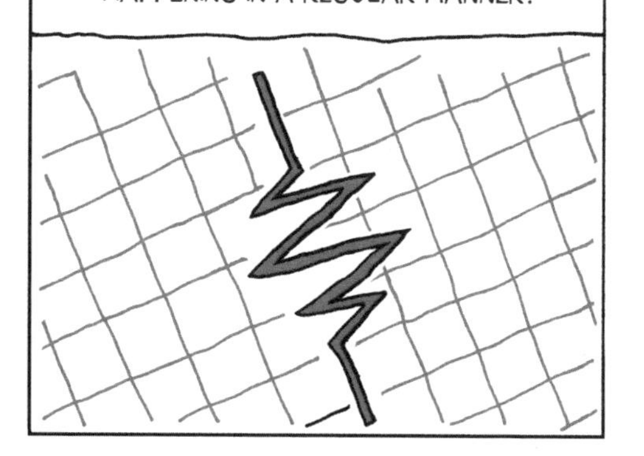
AND SIMPLY CONSIDERED THE SIGNAL AS INTERFERENCE. BUT SHE DIDN'T DO THAT. INSTEAD SHE RECOGNISED THAT SOMETHING WAS HAPPENING, AND HAPPENING IN A REGULAR MANNER.

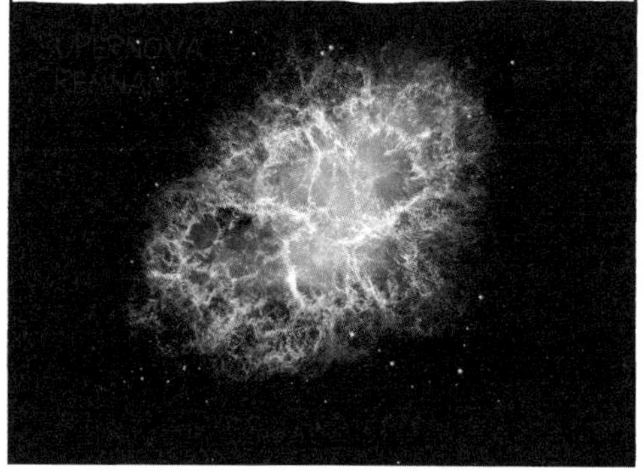
THEN, WITH A DOGGED CONSCIENTIOUSNESS, SHE INVESTIGATED A MYSTERY THAT MANY WOULD HAVE DISMISSED AS UNIMPORTANT.

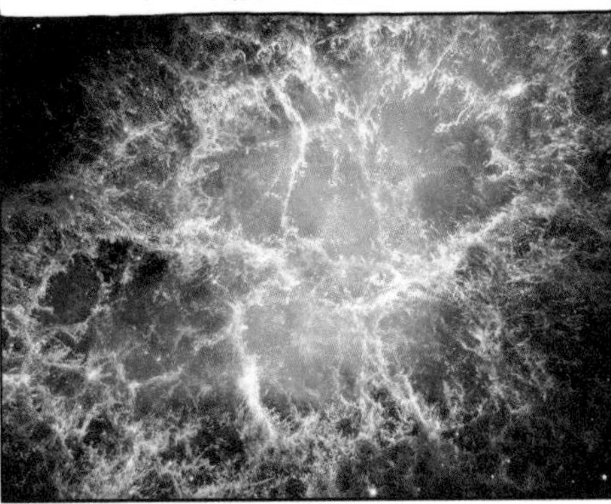
BELL BURNELL KEPT MORE METICULOUS RECORDS THAN HER SUPERVISOR MIGHT HAVE EXPECTED.

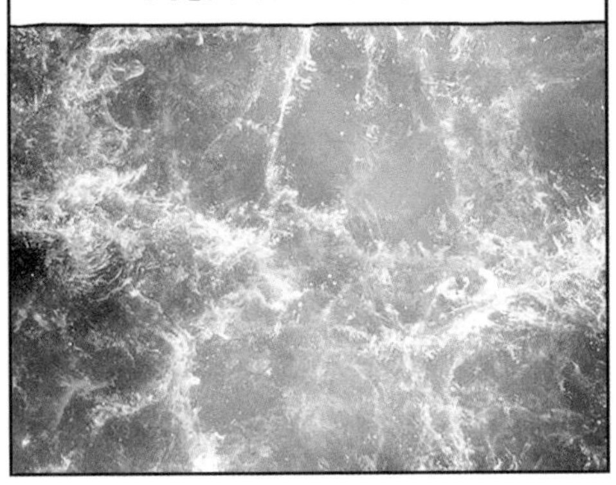
THIS DATA PROVIDED THE CORE EVIDENCE THAT DROVE THE DISCOVERY OF PULSARS.

IF SHE HADN'T DONE THIS IT MIGHT HAVE BEEN YEARS BEFORE ANYONE LOOKED AT THESE SIGNALS SERIOUSLY.

FOR BELL BURNELL, NOT BEING AWARDED THE NOBEL PRIZE DID HER CAREER NO HARM.

SHE WENT ON TO TEACH AT THE UNIVERSITIES OF SOUTHAMPTON, OXFORD, BATH AND PRINCETON. SHE ALSO SERVED AS PROFESSOR OF PHYSICS AT THE OPEN UNIVERSITY FOR TEN YEARS.

SHE WAS PRESIDENT OF THE ROYAL ASTRONOMICAL SOCIETY FROM 2000 TO 2004. SHE WAS AWARDED A CBE IN 1999, WHICH WAS ELEVATED TO A DBE IN 2007.

DAME JOCELYN BELL BURNELL HAD THIS TO SAY ABOUT THE NOBEL PRIZE CONTROVERSY...

THERE ARE SEVERAL COMMENTS THAT I WOULD LIKE TO MAKE ON THIS: FIRST, DEMARCATION DISPUTES BETWEEN SUPERVISOR AND STUDENT ARE ALWAYS DIFFICULT, PROBABLY IMPOSSIBLE TO RESOLVE. SECONDLY, IT IS THE SUPERVISOR WHO HAS THE FINAL RESPONSIBILITY FOR THE SUCCESS OR FAILURE OF THE PROJECT.

WE HEAR OF CASES WHERE A SUPERVISOR BLAMES HIS STUDENT FOR A FAILURE, BUT WE KNOW IT IS LARGELY THE FAULT OF THE SUPERVISOR. IT SEEMS ONLY FAIR TO ME THAT HE SHOULD BENEFIT FROM THE SUCCESS TOO. THIRDLY, I BELIEVE IT WOULD DEMEAN THE NOBEL PRIZES IF THEY WERE AWARDED TO RESEARCH STUDENTS, EXCEPT IN VERY EXCEPTIONAL CASES, AND I DO NOT BELIEVE THIS IS ONE OF THEM. FINALLY, I AM NOT MYSELF UPSET ABOUT IT – AFTER ALL, I AM IN GOOD COMPANY, AM I NOT?

FRED
HOYLE

FRED HOYLE WAS BORN IN GILSTEAD, NEAR BINGLEY, IN YORKSHIRE, ENGLAND, IN 1915.

HIS FATHER, BEN HOYLE, JOINED THE MACHINE GUN CORPS IN THE FIRST WORLD WAR, AND SURVIVED TO RESUME HIS WORK IN THE TEXTILE INDUSTRY ON HIS RETURN.
BRITONS
"WANTS
YOU"
JOIN YOUR COUNTRY'S ARMY!
GOD SAVE THE KING
Reproduced by permission of LONDON OPINION


FRED HOYLE'S MOTHER, MABEL, WHO HAD STUDIED MUSIC AT THE ROYAL ACADEMY, HAD SUPPLEMENTED HER MEAGRE GOVERNMENT ALLOWANCE DURING THE WAR BY PLAYING THE PIANO IN THE LOCAL CINEMA, ACCOMPANYING SILENT MOVIES.

YOU'RE FIRED. WE DON'T NEED FANCY BEETHOVEN PLAYED HERE.

CINEMA ATTENDANCE THEN PLUMMETED. MOST CUSTOMERS ONLY CAME TO HEAR MRS HOYLE PLAY.

OKAY, YOU'RE REHIRED.

IN 1921, MABEL GAVE BIRTH TO A DAUGHTER, JOAN, AND SUFFERED A SERIOUS ILLNESS (POSSIBLY POSTNATAL DEPRESSION).

FRED'S EDUCATION WAS PATCHY IN HIS EARLY YEARS DUE TO HIS FREQUENT TRUANCY.

HE MUCH PREFERRED TO WANDER THE FIELDS AND WOODS AROUND BINGLEY, AND WATCH THE BARGES ON THE LEEDS-LIVERPOOL CANAL...

THAN SUFFER THE ARBITRARY DISCIPLINE OF SCHOOL LIFE, WHICH BORED HIM ANYWAY.

HOYLE HAD A DISTASTE FOR RULES AND POLITE SOCIAL NORMS THROUGHOUT HIS LIFE. NEVERTHELESS, HE DID EVENTUALLY APPLY HIMSELF TO HIS EDUCATION, AND GAINED A PLACE AT BINGLEY GRAMMAR SCHOOL.

FROM THERE HE WON A SCHOLARSHIP TO EMMANUEL COLLEGE, CAMBRIDGE, WHERE HE STUDIED MATHEMATICS. HERE HE EXCELLED AND WON THE MAYHEW PRIZE, WHICH IS GIVEN ANNUALLY TO THE BEST STUDENT OF MATHEMATICS.

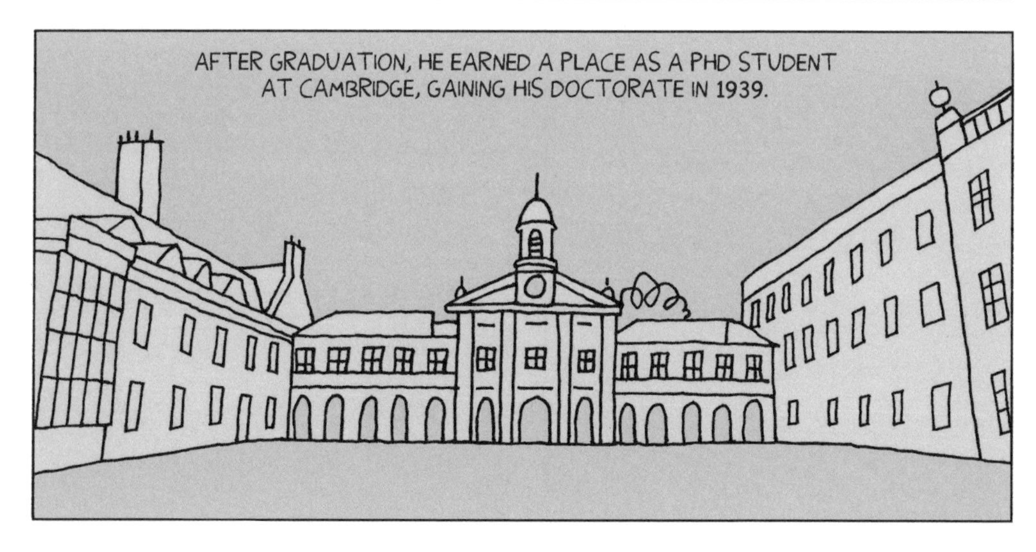
AFTER GRADUATION, HE EARNED A PLACE AS A PHD STUDENT AT CAMBRIDGE, GAINING HIS DOCTORATE IN 1939.

LATER THAT SAME YEAR, HOYLE MARRIED BARBARA CLARK. THE COUPLE WERE TO HAVE TWO CHILDREN – GEOFFREY AND ELIZABETH.

BUT, BEFORE THAT, WAR BROKE OUT.

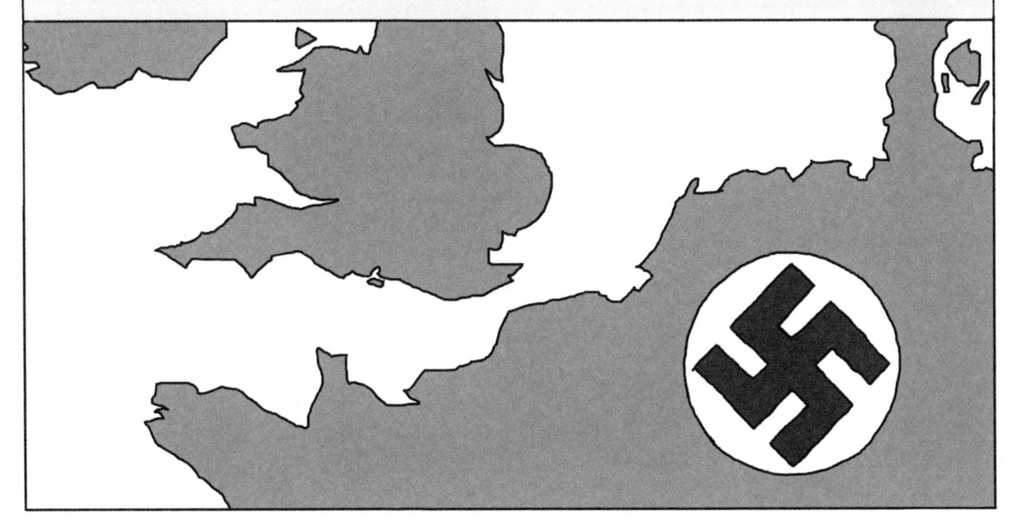
HOYLE LEFT CAMBRIDGE TO GO TO PORTSMOUTH TO WORK FOR THE ADMIRALTY ON RADAR RESEARCH. IT WAS HERE, WHEN STUDYING THE TYPE 79 SHIP-TO-AIRCRAFT EARLY WARNING SYSTEM, THAT HE NOTICED THAT THE DESIGNER HAD NOT FORESEEN THE NEED TO MEASURE THE ALTITUDE OF INCOMING AIRCRAFT, AS WELL AS THE DISTANCE. THIS WAS CRUCIAL INFORMATION IF FIGHTER PLANES WERE TO INTERCEPT ENEMY BOMBERS.

BETWEEN DECEMBER 1940 AND OCTOBER 1941, HOYLE CAME UP WITH A METHOD OF FINDING BOTH DISTANCE AND ALTITUDE...

AND HE DID THIS USING EXISTING EQUIPMENT.

HOYLE NEVER RECEIVED THE CREDIT FOR THIS CRUCIAL BREAKTHROUGH. RADAR WORK DURING THE WAR WAS CLOAKED IN SECRECY.

THE MOST IMPORTANT PERSON IS FRED HOYLE...
HOYLE'S WARTIME COLLEAGUE, THE MATHEMATICIAN HERMANN BONDI

WHO, THROUGH MAKING SIMPLE GRAPHS VERY EARLY ON, ENABLED RADAR OPERATORS ON (AIRCRAFT) CARRIERS TO DETERMINE THE HEIGHT OF A PLANE FROM ITS FADING PATTERN.

I UNDERSTAND THIS WAS OPERATIONALLY VERY SIGNIFICANT, BUT HE HAS NEVER RECEIVED PROPER RECOGNITION BECAUSE IT WAS A SOFTWARE NOT A HARDWARE PROBLEM.

HOYLE STRUCK UP FRIENDSHIPS WITH BOTH HERMANN BONDI AND THE ENGINEER THOMAS GOLD.

IN THE EVENINGS, AFTER A DAY SPENT WITH RADAR EQUIPMENT, THE THREE MEN WOULD TALK ABOUT ASTROPHYSICS AND COSMOLOGY.

THESE LINES OF THOUGHT WERE FURTHER ENCOURAGED ON A TRIP TO THE UNITED STATES MADE BY HOYLE AS PART OF HIS WAR WORK.

IN PASADENA, AT THE MOUNT WILSON OBSERVATORY, HE MET THE EMINENT ASTRONOMER WALTER BAADE.

BAADE, A GERMAN NATIONAL, WAS CONFINED TO THE LOS ANGELES AREA.

HOWEVER, BEING A PROMINENT SCIENTIST, HE WAS NOT SUBJECT TO THE CURFEWS BY WHICH OTHER ENEMY ALIENS WERE BOUND DURING THE WAR.

BAADE, WHO HAD ACCESS TO THE MOST POWERFUL TELESCOPE IN THE WORLD, HAD TAKEN ADVANTAGE OF THE DARK SKIES CAUSED BY THE LOS ANGELES BLACKOUTS TO SURVEY THE ANDROMEDA GALAXY. HE WAS THE FIRST PERSON TO RESOLVE THE AMORPHOUS BLUR INTO INDIVIDUAL STARS. THIS WORK LED ON TO HIS DISCOVERY THAT THERE WERE TWO DISTINCT POPULATIONS OF STARS IN THE UNIVERSE.

THE FIRST POPULATION, WHICH INCLUDES OUR OWN SUN, IS COMMON IN THE SPIRAL ARMS OF GALAXIES. THE BRIGHTEST OF THESE ARE BLUE AND RED SUPERGIANTS.

THE SECOND ARE FOUND IN ELLIPTICAL GALAXIES, THE CORES OF SPIRAL GALAXIES, AND GLOBULAR CLUSTERS.

HOYLE IMMEDIATELY SAW THE IMPLICATIONS OF BAADE'S FINDINGS. WHAT MECHANISMS, HE WONDERED, WERE AT WORK TO SORT THE STARS INTO SUCH DIFFERENT POPULATIONS?

WHY DID ONE POPULATION NOT HAVE ANY RED GIANT STARS? WHY DID THE POPULATIONS OF LARGE GLOBULAR CLUSTERS AND SMALL ELLIPTICAL GALAXIES LOOK SO SIMILAR?

AFTER THE WAR, HOYLE, GOLD AND BONDI ALL RETURNED TO CAMBRIDGE TO RESUME THEIR ACADEMIC CAREERS. THEIR THOUGHTS INCREASINGLY TURNED TO THE NATURE OF THE UNIVERSE.

IN THE 1920s THE AMERICAN ASTRONOMER EDWIN HUBBLE HAD MADE TWO MAJOR DISCOVERIES. AT THE TIME THE PREVAILING VIEW WAS THAT THE ENTIRE UNIVERSE WAS COMPOSED OF JUST OUR MILKY WAY GALAXY.

HUBBLE PROVED THAT THE NEBULAS (INDISTINCT BRIGHT PATCHES) SEEN BEYOND OUR MILKY WAY WERE MUCH TOO DISTANT TO BE PART OF OUR GALAXY, BUT WERE, IN FACT, ENTIRE GALAXIES OUTSIDE OUR OWN.

HUBBLE ALSO CONFIRMED THAT THESE GALAXIES (REDSHIFTED, DUE TO THEIR LIGHT BEING STRETCHED AS THEY MOVED AWAY)...

WERE RECEDING AT GREAT VELOCITY. THE UNIVERSE WAS NOT STATIC, BUT EXPANDING.

THE COSMOS WAS GROWING IN SIZE, WITH ITS GALAXIES FLYING EVER FURTHER AWAY FROM EACH OTHER.

HUBBLE'S FINDINGS SUGGESTED THAT IN THE DISTANT PAST THE UNIVERSE WAS MUCH SMALLER...

AND HAD ULTIMATELY STARTED OUT AS A SINGLE POINT.

MANY SCIENTISTS WERE UNEASY WITH THE CONCEPT OF A UNIVERSE THAT BEGAN IN AN EXPLOSION THAT CAME FROM NOTHING.

HOYLE WOULD LATER COMPARE THE UNIVERSE BURSTING OUT OF NOWHERE TO A 'PARTY GIRL JUMPING OUT OF A BIRTHDAY CAKE'.

IT JUST WASN'T DIGNIFIED OR ELEGANT.

IN 1946, HOYLE, ALONG WITH GOLD AND BONDI, SAW THE EALING FILM *DEAD OF NIGHT*.
EXIT

THIS MOVIE, A HORROR FILM, WAS VERY DIFFERENT FROM THE GENTEEL ENGLISH COMEDIES THAT EALING STUDIOS HAD BEEN PRODUCING UP TO THIS TIME.

IN THE STORY, A MAN IS INVITED TO A COUNTRY HOUSE, WHERE HE TELLS THE ASSEMBLED GROUP OF PEOPLE, THAT HE'S MET THEM ALL BEFORE IN A RECURRING DREAM.

FASCINATED, THE OTHER GUESTS THEN TELL THEIR OWN EXPERIENCES OF THE SUPERNATURAL – FIVE TALES INVOLVING PREMONITIONS, GHOSTS, A HAUNTED MIRROR AND A VENTRILOQUIST'S DUMMY.

THE FILM ENDS WITH A SURREAL SEQUENCE OF INCREASING HORROR...

THAT CULMINATES WITH THE MAIN CHARACTER WAKING UP IN FRIGHT.

HE IS THEN INVITED BY TELEPHONE TO A SMALL COTTAGE IN THE COUNTRY...

AND AS THE CREDITS ROLL, THE WHOLE STORY BEGINS AGAIN.

*DEAD OF NIGHT* HAD A STRONG EFFECT ON THE THREE MEN. THOMAS GOLD MUSED THAT PERHAPS THE UNIVERSE WAS STRUCTURED LIKE THE FILM...
CINEMA

WITH MUCH ACTIVITY, BUT NO BEGINNING AND NO END? AN ETERNAL AND INFINITE COSMOS.

GOLD WORKED ON THE IDEA. OBSERVATIONS SHOWED THAT THE UNIVERSE WAS EXPANDING. IF THIS WAS TRUE THEN YOU WOULD EXPECT THE UNIVERSE TO BECOME LESS DENSE AS GALAXIES MOVED AWAY FROM ONE ANOTHER.

BUT WHAT IF THE UNIVERSE COMPENSATED FOR THIS BY CREATING MATTER IN THE GROWING GAPS? SUCH A UNIVERSE WOULD DEVELOP AND EXPAND, YET REMAIN ESSENTIALLY UNCHANGED.

THE 'STEADY STATE MODEL', AS IT BECAME KNOWN, DISPENSED WITH THE CLUMSY EXPLOSIVE BEGINNING TO THE COSMOS, AND REPLACED IT WITH AN INFINITE UNIVERSE, CONSTANT AND ETERNAL, THAT WOULD BE REPLENISHED AS IT WAS DEPLETED BY EXPANSION.

WHERE, THEN, WAS THIS MATTER BEING CREATED AND WHAT WAS IT COMING FROM? HOYLE CALCULATED THAT COMPENSATION FOR THE UNIVERSE'S EXPANSION REQUIRED A CREATION RATE OF ONLY...

ONE ATOM EVERY CENTURY IN A VOLUME EQUAL TO THAT OF THE EMPIRE STATE BUILDING.

TO EXPLAIN THE EXISTENCE OF THESE ATOMS, HOYLE PROPOSED THE CONCEPT OF THE CREATION FIELD (C-FIELD).

THE C-FIELD WAS SUPPOSED TO PERMEATE THE ENTIRE UNIVERSE, GENERATING ATOMS, AND MAINTAINING THE STATUS QUO.

HOYLE ADMITTED THAT HE HAD NO IDEA HOW THE C-FIELD ACTUALLY WORKED.

YES, IT WAS A STRANGE IDEA, BUT WAS IT ANY STRANGER THAN THE THEORY THAT EVERYTHING HAD APPEARED ALL AT ONCE OUT OF A MIGHTY EXPLOSION?

NO ONE COULD REALLY EXPLAIN THAT EITHER.

THE CAMBRIDGE TRIO DEVELOPED GOLD'S IDEAS FURTHER IN TWO PAPERS PUBLISHED IN 1947.

BOTH OF THESE PAPERS HAD AN ENORMOUS IMPACT ON COSMOLOGY, SPLITTING THE FIELD INTO TWO FACTIONS, AND CAUSING A BATTLE THAT WOULD LAST TWENTY YEARS.

HOYLE HIMSELF INADVERTENTLY COINED THE POPULAR TERM FOR THE THEORY THAT OPPOSED HIS STEADY STATE MODEL...

IN A BROADCAST ON BBC RADIO'S *THIRD PROGRAMME* IN 1949.

NOW THIS BIG BANG THEORY IDEA SEEMS TO ME TO BE UNSATISFACTORY.

HE MAY HAVE INTENDED THE PHRASE AS DERISORY. NEVERTHELESS, IT WAS GRADUALLY ADOPTED.

THE REGULAR RADIO BROADCASTS HE MADE BECAME A SIDE CAREER FOR HOYLE. THE UNDERGRADUATES HE TAUGHT AT CAMBRIDGE WOULD OFTEN ARRIVE FOR SUPERVISION...

ONLY TO FIND...

THE FIRST SERIES OF LECTURES HE RECORDED FOR THE BBC WAS COLLECTED TOGETHER INTO A BOOK.
THE NATURE OF THE UNIVERSE
FRED HOYLE

HOYLE WOULD GO ON TO WRITE MANY OTHER POPULAR SCIENCE BOOKS ON SUBJECTS AS DIVERSE AS ASTRONOMY, NUCLEAR ENERGY, AND STONEHENGE.

HE ALSO BEGAN TO WRITE SCIENCE FICTION. HIS FIRST NOVEL, *THE BLACK CLOUD*, WAS PUBLISHED IN 1957.

THE STORY RELATES HOW A VAST CLOUD OF INTERSTELLAR GAS ENTERS THE SOLAR SYSTEM, BLOCKING THE SUN'S RADIATION.

DARKNESS FALLS ACROSS THE EARTH, TEMPERATURES PLUMMET, CROPS FAIL, AND PEOPLE STARVE.

IT IS DISCOVERED THAT THE CLOUD IS AN ALIEN GASEOUS SUPER-ORGANISM, MANY TIMES MORE INTELLIGENT THAN HUMANS.

THE CLOUD IS SURPRISED TO DISCOVER INTELLIGENT LIFE ON A SOLID PLANET. WHEN ASKED HOW ITS OWN LIFE FORM ORIGINATED, THE ALIEN REPLIES THAT IT HAS ALWAYS BEEN THERE.

ULTIMATELY, AFTER RECEIVING INFORMATION THAT ANOTHER OF ITS KIND HAS VANISHED IN NEARBY SPACE, THE CLOUD RETREATS FROM THE SOLAR SYSTEM TO INVESTIGATE, RELIEVING EARTH'S SUFFERING.

FRED HOYLE WENT ON TO WRITE MORE SCIENCE FICTION NOVELS, OFTEN IN COLLABORATION WITH HIS SON, GEOFFREY.

IN 1961, HOYLE, ALONG WITH TELEVISION PRODUCER JOHN ELLIOT, WROTE A SEVEN-PART TV SERIES FOR THE BBC.

*A FOR ANDROMEDA* CONCERNS A GROUP OF SCIENTISTS WHO DETECT A RADIO SIGNAL FROM DEEP SPACE...

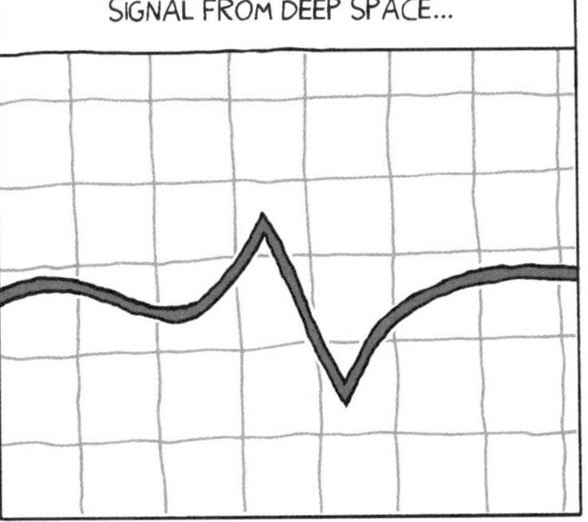

THAT CONTAINS INSTRUCTIONS FOR THE DESIGN OF AN ADVANCED COMPUTER.

WHEN THE COMPUTER IS BUILT, IT GIVES THE SCIENTISTS INSTRUCTIONS TO BUILD A LIVING ORGANISM...

WHICH THEY NAME 'ANDROMEDA' (JULIE CHRISTIE IN HER FIRST MAJOR ROLE). GRADUALLY IT IS REVEALED THAT THE ALIENS HAVE HOSTILE INTENT. IN THE END, THEY ARE DEFEATED, AND ANDROMEDA IS DESTROYED.

HOYLE'S CREATIVE INTERESTS WERE WIDE-RANGING. HE WROTE NOT ONLY FICTION, NON-FICTION AND RADIO LECTURES, BUT ALSO A THEATRE PLAY AND CHILDREN'S BOOKS, ALL PROMOTING SCIENCE. YET, DESPITE ALL THIS OTHER WORK, HE DID NOT NEGLECT HIS INVESTIGATIONS INTO COSMOLOGY. SINCE THE 1940s HE'D BEEN INTERESTED IN STELLAR NUCLEOSYNTHESIS – THE PROCESS BY WHICH STARS CREATE HEAVY ELEMENTS, LIKE IRON, FROM LIGHTER ELEMENTS, LIKE HYDROGEN.

THE MAIN PROPONENT OF THE BIG BANG THEORY WAS GEORGE GAMOW, A RUSSIAN PHYSICIST WHO HAD DEFECTED TO THE UNITED STATES.

HE BELIEVED THAT ALL THE VARIOUS ATOMIC NUCLEI THAT MADE UP THE UNIVERSE WERE CREATED IN THE HOUR IMMEDIATELY FOLLOWING THE BIG BANG.

HOWEVER, DESPITE ALL OF GAMOW AND HIS TEAM'S BEST EFFORTS, NO MECHANISM THAT WOULD CREATE ANYTHING BUT THE LIGHTEST ELEMENTS COULD BE FOUND.

IT SEEMED THAT THE EARLY UNIVERSE HAD BEEN MADE OF MOSTLY HYDROGEN AND HELIUM.

IF THIS WAS SO, THEN WHERE DID ALL THE HEAVIER ELEMENTS COME FROM?

IT WAS HOYLE WHO CONTRIBUTED THE MOST TO SOLVING THIS PROBLEM.

FOR HIM IT WASN'T A 'STEADY STATE VERSUS BIG BANG' PROBLEM, BUT AN ISSUE THAT CONCERNED BOTH MODELS.

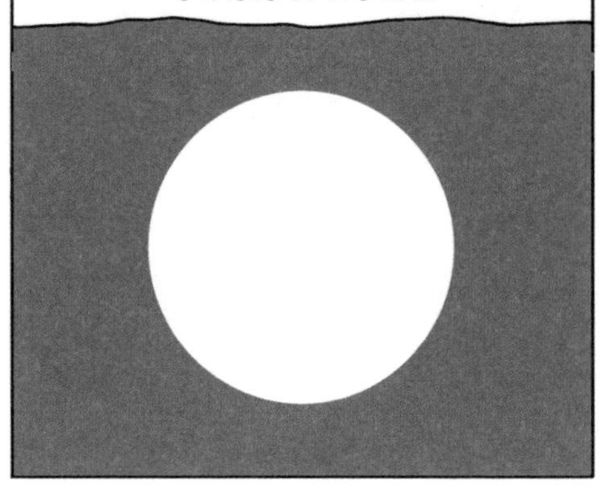
SO HOYLE BEGAN TO THINK ABOUT WHAT WOULD HAPPEN TO A STAR AS IT PASSED THROUGH THE VARIOUS STAGES OF ITS LIFE.

IN PARTICULAR, WHAT WOULD HAPPEN AT THE END OF A STAR'S LIFE WHEN IT HAD BEGUN TO RUN OUT OF HYDROGEN FUEL. THE CALCULATIONS HE MADE WERE QUITE CLEAR. A FUEL SHORTAGE WOULD LEAD TO A FALL IN OUTWARD PRESSURE. THE STAR WOULD BEGIN TO CONTRACT UNDER GRAVITATIONAL FORCE. AS THE STAR FALLS INWARD, THE STELLAR CORE BEGINS TO HEAT UP, CAUSING FURTHER NUCLEAR REACTIONS, RENEWING OUTWARD PRESSURE, AND HALTING THE PROCESS OF COLLAPSE.

THIS STABILITY IS ONLY TEMPORARY. THE STAR CONTINUES TO BURN FUEL AND, AS THE CORE COOLS AGAIN, A NEW COLLAPSING PHASE BEGINS.

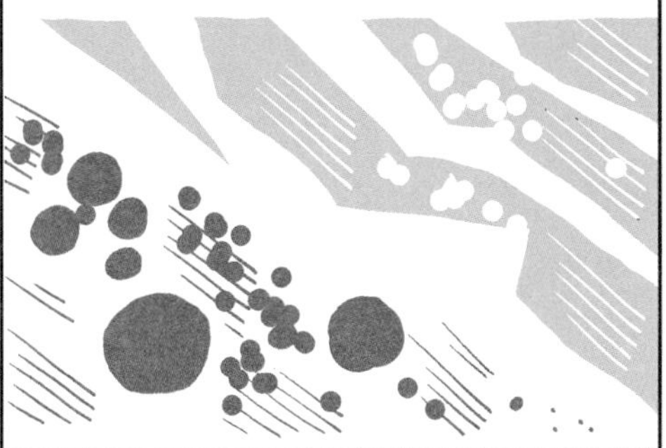

INCREASED PRESSURE CAUSES MORE NUCLEAR REACTIONS, STABILISING THE STAR ONCE MORE, UNTIL THE NEXT FUEL SHORTAGE.

THIS STOP-START COLLAPSING PROCESS CONTINUES, WITH THE STAR SHRINKING FURTHER EACH TIME. HOYLE ANALYSED VARIOUS TYPES OF STAR (SMALL, MEDIUM AND LARGE) AND AFTER SEVERAL YEARS COMPLETED HIS CALCULATIONS OF THE TEMPERATURE AND PRESSURE CHANGES THAT HAPPEN IN A STAR'S FINAL THROES. HE ALSO WORKED OUT THE NUCLEAR REACTIONS IN EACH DEATH SPASM, SHOWING HOW VARIOUS COMBINATIONS OF EXTREME TEMPERATURES AND PRESSURES COULD LEAD TO A WHOLE RANGE OF MEDIUM-WEIGHT AND HEAVYWEIGHT ATOMIC NUCLEI.

HOYLE'S CALCULATIONS COULD EVEN ACCOUNT FOR THE EXACT QUANTITIES OF ALMOST ALL THE ELEMENTS WE SEE TODAY...

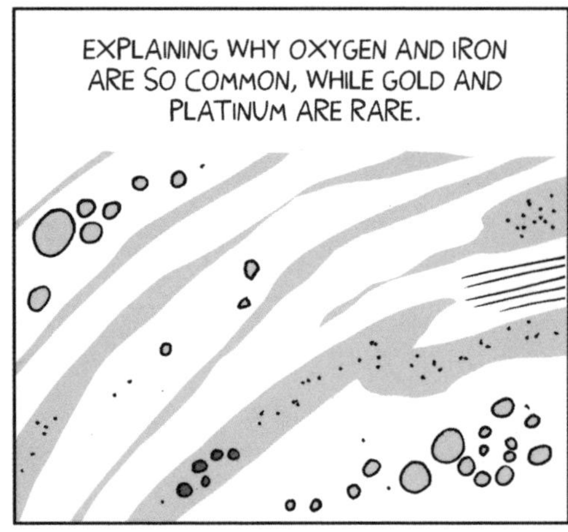
EXPLAINING WHY OXYGEN AND IRON ARE SO COMMON, WHILE GOLD AND PLATINUM ARE RARE.

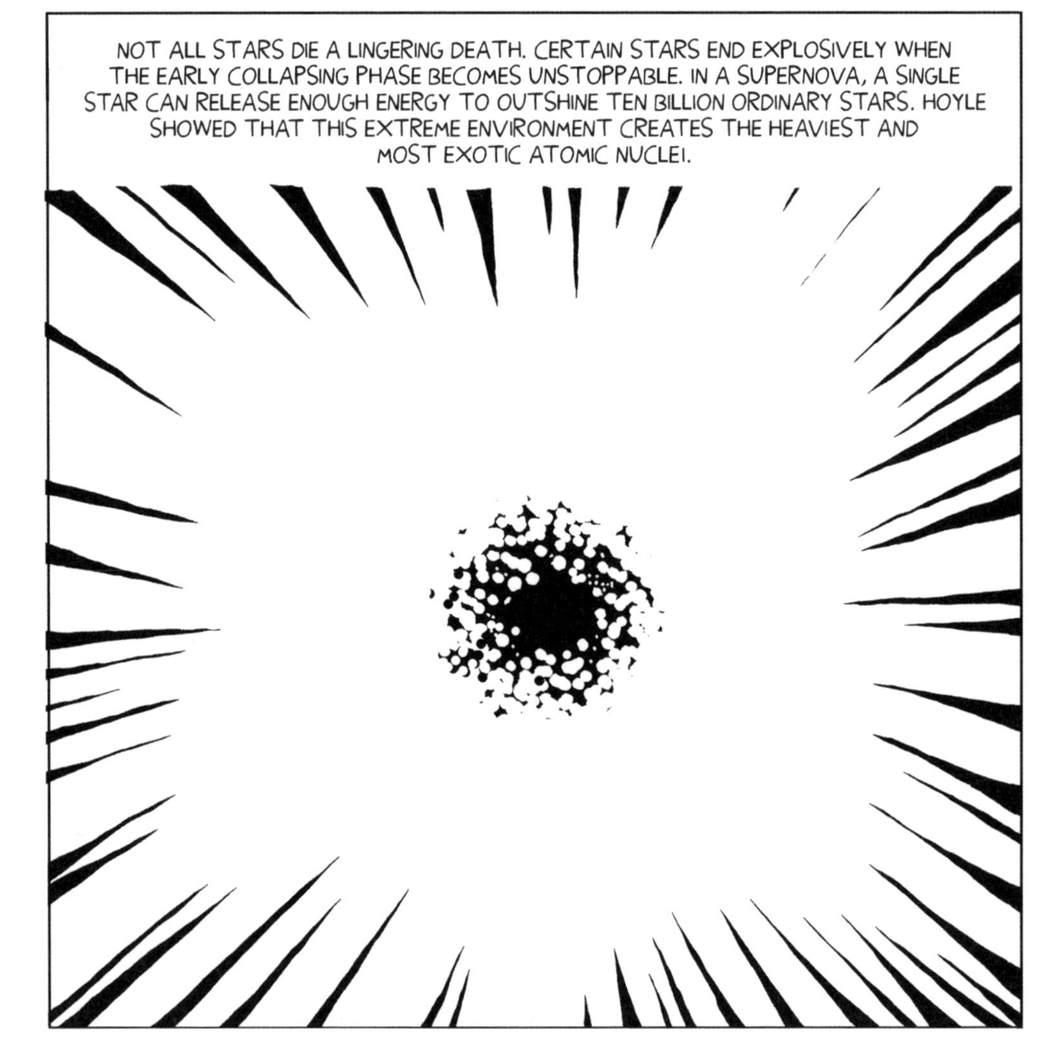
NOT ALL STARS DIE A LINGERING DEATH. CERTAIN STARS END EXPLOSIVELY WHEN THE EARLY COLLAPSING PHASE BECOMES UNSTOPPABLE. IN A SUPERNOVA, A SINGLE STAR CAN RELEASE ENOUGH ENERGY TO OUTSHINE TEN BILLION ORDINARY STARS. HOYLE SHOWED THAT THIS EXTREME ENVIRONMENT CREATES THE HEAVIEST AND MOST EXOTIC ATOMIC NUCLEI.

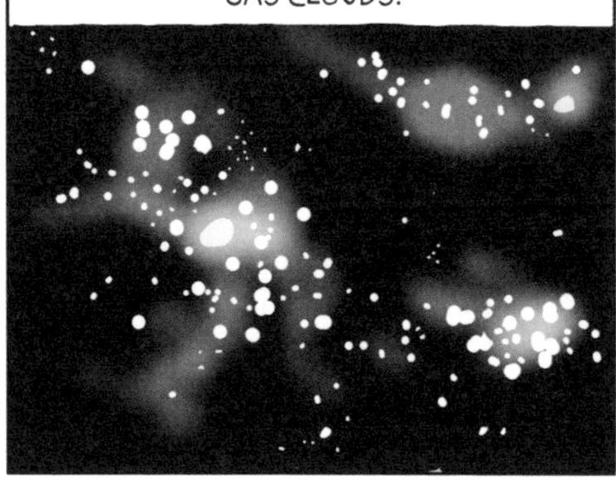

DEBRIS FROM A SUPERNOVA IS BLASTED ACROSS THE UNIVERSE IN VAST GAS CLOUDS.

THIS MATERIAL EVENTUALLY CONDENSES TO FORM NEW STARS...

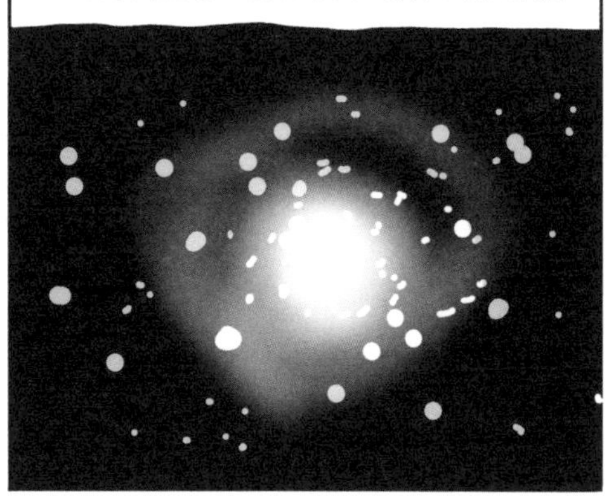

WHICH WILL HAVE A HEAD START IN TERMS OF NUCLEOSYNTHESIS, BECAUSE THEY ALREADY CONTAIN HEAVY ATOMS.

SO WHEN THESE STARS DIE AND EXPLODE...

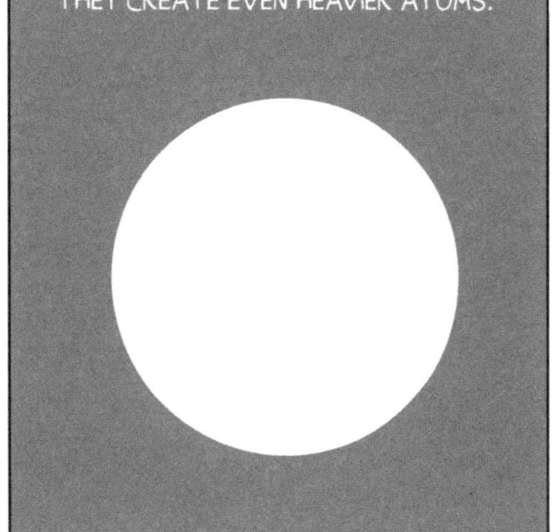

THEY CREATE EVEN HEAVIER ATOMS.

IT IS THOUGHT THAT OUR OWN SUN IS A THIRD-GENERATION STAR.

THE ELEMENTS THAT MAKE UP THE ROCK AND SOIL UNDER OUR FEET...

THE IRON IN OUR BLOOD...

THE CALCIUM IN OUR BONES...

THE OXYGEN THAT FILLS OUR LUNGS...

OUR BODIES AND THE BODIES OF ALL LIVING THINGS...

WERE FORGED IN LONG-DEAD STARS OVER BILLIONS OF YEARS.

THE RICHNESS OF LIFE WOULDN'T EXIST WITHOUT THIS PROCESS.

YET HOYLE HAD ONE PROBLEM TO RESOLVE BEFORE HE COULD SAY THAT HE'D WORKED OUT THE COMPLETE CHAIN OF ELEMENT-BUILDING, WHICH STARTED WITH HYDROGEN AND HELIUM, AND ENDED WITH ALL THE HEAVIER ATOMS. THE SPECIFICS OF HOW HELIUM BECAME CARBON DEFEATED HIM.

THERE WAS NO VIABLE PATHWAY THAT HE COULD SEE, AND, IF HE COULD NOT EXPLAIN HOW CARBON WAS CREATED, THEN HE COULD NOT SAY HOW ALL THE OTHER NUCLEAR REACTIONS TOOK PLACE, AS THEY ALL NEEDED CARBON AT SOME POINT.

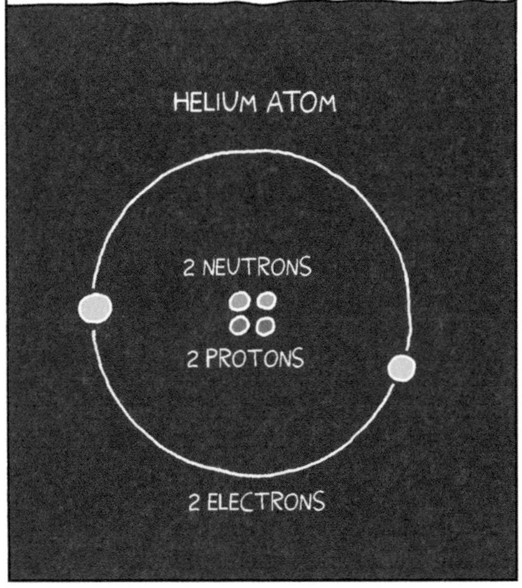

HOYLE GAVE THOUGHT TO HOW THE SIMPLE HELIUM NUCLEUS COULD BECOME A MORE COMPLEX CARBON NUCLEUS.

CARBON ATOM

6 ELECTRONS

6 PROTONS

6 NEUTRONS

WHAT IF THREE HELIUM NUCLEI WERE TO COLLIDE SIMULTANEOUSLY TO CREATE CARBON? THIS SEEMED TO HIM TO BE EXTREMELY UNLIKELY.

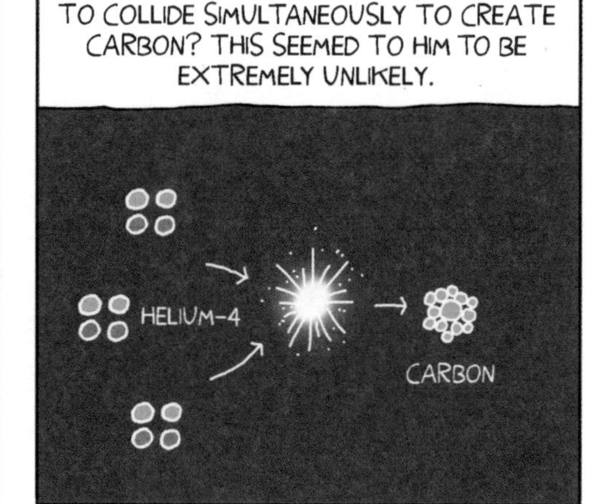


BUT WHAT IF ONLY TWO HELIUM NUCLEI COLLIDED? THIS WOULD MAKE A BERYLLIUM-8 NUCLEUS.

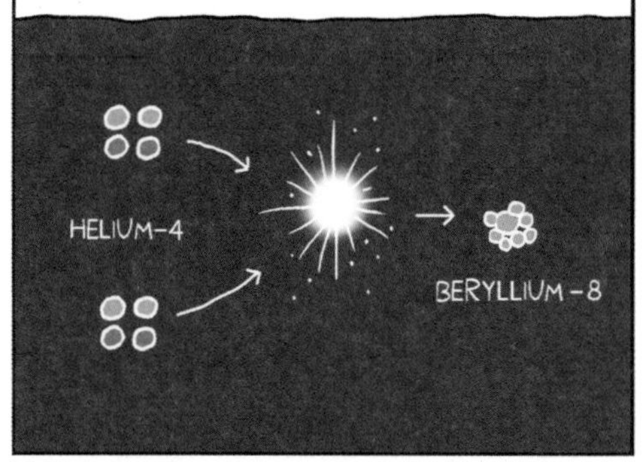


IF THIS BERYLLIUM-8 WENT ON AND FUSED WITH A THIRD HELIUM NUCLEUS, THEN THE RESULT WOULD BE CARBON.

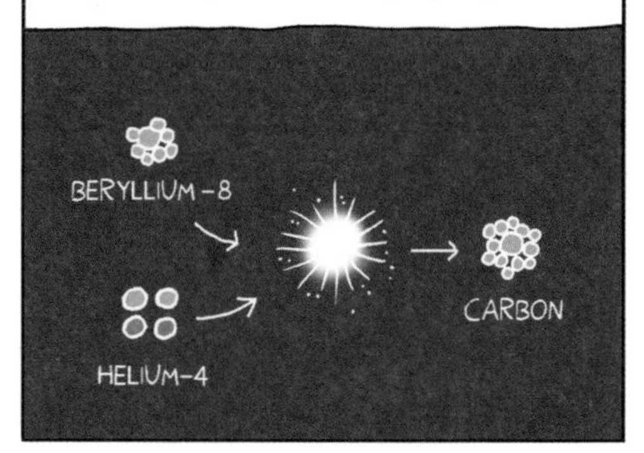


THE PROBLEM HERE IS THAT BERYLLIUM-8 IS VERY UNSTABLE. IT TYPICALLY LASTS FOR LESS THAN A MILLIONTH OF A BILLIONTH OF A SECOND BEFORE DISINTEGRATING.

THERE WAS YET A FURTHER DIFFICULTY. THE COMBINED MASS OF A BERYLLIUM-8 AND A HELIUM NUCLEUS IS FRACTIONALLY MORE THAN THAT OF A CARBON NUCLEUS. SO WHERE DID THE EXTRA MASS GO?

THE MASS COULDN'T BE CONVERTED INTO ENERGY, BECAUSE A BERYLLIUM NUCLEUS WOULDN'T BE AROUND LONG ENOUGH FOR THE PROCESS TO TAKE PLACE.

BUT PERHAPS THIS ENERGY DIDN'T GO ANYWHERE. HOYLE CONCLUDED THAT THERE MUST BE A FORM OF CARBON WITH EXACTLY THE SAME MASS AS THE COMBINED BERYLLIUM-8 AND HELIUM-4...
$^{4}He + ^{4}He \rightarrow ^{8}Be + Y$
$^{4}He + ^{8}Be \rightarrow ^{12}C + Y$

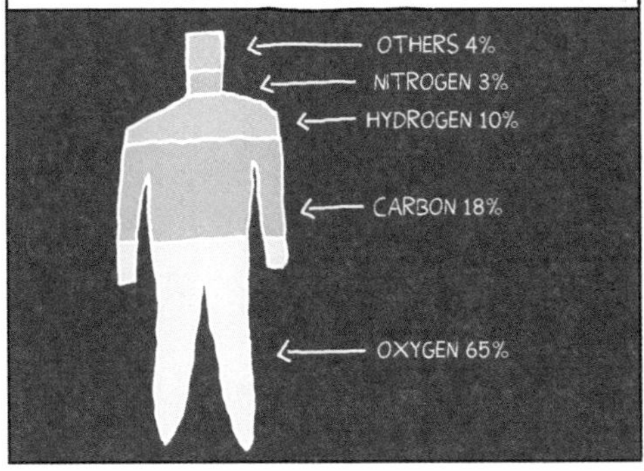
WITH 7.65 MEGAELECTRONVOLTS MORE ENERGY THAN THE BASIC CARBON NUCLEUS. HE REASONED THAT, AS HE WAS PARTIALLY MADE FROM CARBON-12...
OTHERS 4%
NITROGEN 3%
HYDROGEN 10%
CARBON 18%
OXYGEN 65%

THEN THIS SECOND STATE OF CARBON MUST EXIST, OTHERWISE, HE WOULDN'T EXIST HIMSELF.
HEH!

WE ARE HERE TO STUDY THE UNIVERSE, SO THE LAWS OF THE UNIVERSE MUST BE COMPATIBLE WITH OUR EXISTENCE.

HE WAS SATISFIED IN HIS OWN MIND THAT HE'D SOLVED THE PROBLEM. NOW ALL HE HAD TO DO WAS FIND THE EVIDENCE TO CONVINCE OTHERS.

HOYLE'S SABBATICAL LEAVE TO THE CALIFORNIA INSTITUTE OF TECHNOLOGY (CALTECH) IN 1953 GAVE HIM THE CHANCE HE NEEDED.

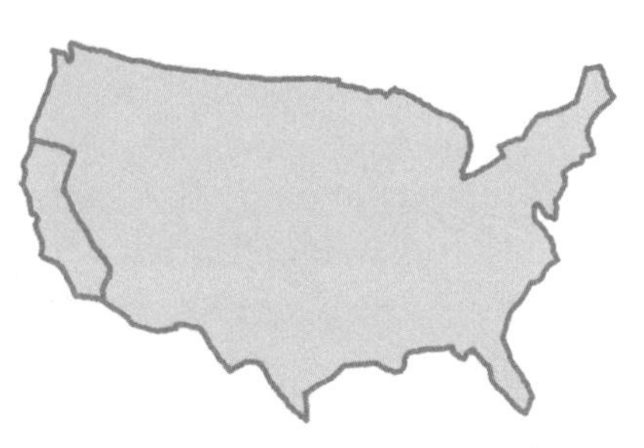

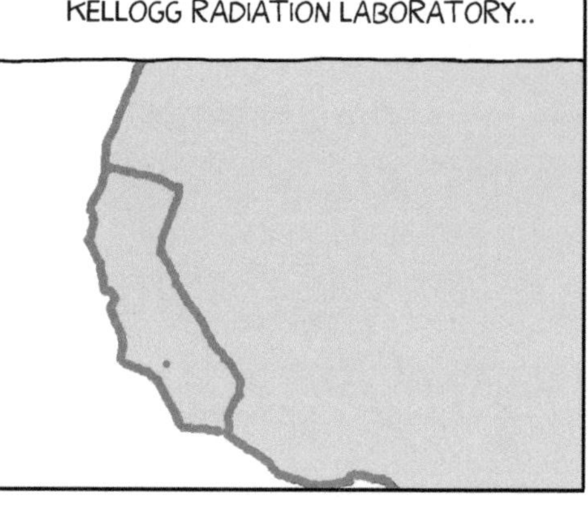


WHERE THE AMERICAN NUCLEAR PHYSICIST WILLIAM FOWLER AND HIS TEAM WERE USING A PARTICLE ACCELERATOR (A MACHINE THAT USES ELECTROMAGNETIC FIELDS TO PROPEL CHARGED PARTICLES TO NEARLY LIGHT SPEEDS) TO EXPLORE NUCLEOSYNTHESIS.

I WAS VERY SCEPTICAL THAT THIS STEADY STATE COSMOLOGIST, THIS THEORIST, SHOULD ASK QUESTIONS ABOUT THE CARBON NUCLEUS.

UNDAUNTED, HOYLE ARGUED THAT FOWLER COULD CHECK HIS THEORY EASILY IN JUST A FEW DAYS. IF HOYLE WAS WRONG, THEN FOWLER WOULD ONLY HAVE TO WORK THREE OR FOUR EXTRA NIGHTS TO CATCH UP...

BUT, IF HE WAS RIGHT, THEN HE WOULD HAVE MADE ONE OF THE BIGGEST DISCOVERIES IN NUCLEAR PHYSICS.

FOWLER'S TEAM DID FIND THE EXCITED STATE OF CARBON-12, AND IT WAS AT 7.65 MeV, EXACTLY WHERE HOYLE SAID IT WOULD BE.

IT WAS A STUNNING PREDICTION ON THE PART OF THE ENGLISHMAN.

WILLIAM FOWLER CONTINUED TO WORK WITH HOYLE. IN 1957, THEY, TOGETHER WITH THE HUSBAND-AND-WIFE TEAM OF MARGARET AND GEOFFREY BURBIDGE...

PRODUCED THE INFLUENTIAL PAPER 'SYNTHESIS OF THE ELEMENTS IN STARS', ALSO KNOWN AS THE B2FH PAPER, AFTER THE INITIALS OF ITS FOUR AUTHORS.

THIS PAPER, IN A GENERAL WAY, EXPLAINED THE ABUNDANCE OF PRACTICALLY ALL THE ELEMENTS IN THE UNIVERSE, FROM HYDROGEN TO URANIUM, BY SYNTHESIS IN THE STARS AND SUPERNOVAE.

IT IS WIDELY RECOGNISED AS ONE OF THE GREATEST TRIUMPHS OF 20TH-CENTURY SCIENCE.

IN 1983, WILLIAM FOWLER, ALONG WITH SUBRAHMANYAN CHANDRASEKHAR, WON THE NOBEL PRIZE IN PHYSICS FOR HIS 'THEORETICAL AND EXPERIMENTAL STUDIES OF THE NUCLEAR REACTIONS OF IMPORTANCE IN THE FORMATION OF THE CHEMICAL ELEMENTS IN THE UNIVERSE.'

THE FACT THAT FRED HOYLE DIDN'T RECEIVE THE NOBEL PRIZE AS WELL IS CONSIDERED BY MANY TO BE ONE OF THE GREAT INJUSTICES OF THE AWARD.

FOWLER HIMSELF WAS DEVASTATED WHEN HE REALISED HOYLE HAD BEEN OVERLOOKED.

FOWLER WROTE TO HOYLE: 'AFTER THE INITIAL ELATION AND EXCITEMENT I HAVE HAD A HEAVY HEART FOR TWO WEEKS. IT IS IMPOSSIBLE TO UNDERSTAND WHY THE PRIZE WAS NOT GIVEN TO YOU OR SHARED BETWEEN US. I REALISE THAT NOTHING I CAN WRITE WILL HELP, BUT THIS PERSONAL NOTE TO YOU WILL HELP MY OWN FEELINGS.'

SO WHY DID THE NOBEL PRIZE COMMITTEE OVERLOOK HOYLE? IN MARCH 1975, HOYLE HAD MADE SOME UNFORTUNATE REMARKS IN AN INTERVIEW ABOUT ANTONY HEWISH AND MARTIN RYLE, WHO HAD WON THE NOBEL PRIZE FOR PHYSICS IN 1974 FOR THEIR ROLE IN THE DISCOVERY OF PULSARS.

THERE WAS NOTHING IN THE PROCEEDINGS THAT TOLD ME THAT JOCELYN BELL HAD BEEN THE DISCOVERER. I WAS A FELLOW PROFESSOR WITH HEWISH AT CAMBRIDGE AT THE TIME. THERE WERE ABOUT EIGHT OF THEM AND THE WHOLE THING WAS UNDER WRAPS FOR SIX MONTHS. NO ONE IN THE GROUP WAS ALLOWED TO SPEAK TO ANYONE OUTSIDE. THEY WERE BUSY PINCHING IT FROM THE GIRL. THAT'S WHAT IT AMOUNTED TO.

THESE OFF-THE-CUFF REMARKS MADE HEADLINES AROUND THE WORLD.

FEARING LIBEL ACTION, HOYLE WROTE A CAREFULLY COMPOSED EXPLANATION TO *THE TIMES* OF LONDON. THE FINAL PARAGRAPH OF THIS LETTER READS...

I WOULD ADD THAT MY CRITICISM OF THE NOBEL PRIZE WAS DIRECTED AT THE AWARDS COMMITTEE ITSELF, NOT PROFESSOR HEWISH.

IT SEEMS CLEAR THAT THE COMMITTEE DID NOT BOTHER ITSELF TO UNDERSTAND WHAT HAPPENED IN THIS CASE.

SUCH CRITICISM WOULD NOT HAVE GONE DOWN WELL IN SWEDEN. BUT, APART FROM THIS, THERE ARE OTHER REASONS WHICH CONTRIBUTED TO HOYLE'S MISSING OUT ON THE BIGGEST PRIZE IN PHYSICS.

HIS TENDENCY TO MAKE IMAGINATIVE LEAPS IN LOGIC, COMBINED WITH HIS COMBATIVENESS, LED HIM TO TAKE POSITIONS ON SUBJECTS WAY OUTSIDE ESTABLISHED THOUGHT, AND THE EVIDENCE HE PRESENTED TO SUPPORT THAT VIEW WAS OFTEN THIN AT BEST.

IT WAS AS IF HIS EARLY SUCCESS WITH NUCLEOSYNTHESIS MADE HIM OVERCONFIDENT IN HIS INTUITION, WITH THE RESULT THAT THE IMAGINATIVE IDEAS HE'D PREVIOUSLY RESERVED FOR SCIENCE FICTION GOT PRESENTED AS SCIENCE FACT.

HOYLE PROMOTED THE THEORY OF PANSPERMIA – A CONCEPT HE DEVELOPED WITH HIS FORMER STUDENT, CHANDRA WICKRAMASINGHE. ACCORDING TO THE TWO MEN, LIFE DID NOT ORIGINATE ON EARTH BUT OUT IN SPACE, WHERE IT EXISTS IN ABUNDANCE THROUGHOUT THE UNIVERSE.

IT IS DISTRIBUTED BY COMETS, WHICH CONTAIN MICROBES THAT HAVE EVOLVED FROM ORGANIC COMPOUNDS THAT EXIST WITHIN THESE BODIES. LARGER COMETS WOULD HAVE WARM LIQUID INTERIORS WHERE THE ASSEMBLY OF MOLECULES INTO PRIMITIVE LIVING CELLS COULD TAKE PLACE. EVOLUTION ON EARTH IS INFLUENCED BY A STEADY FALL OF VIRUSES FROM SPACE.

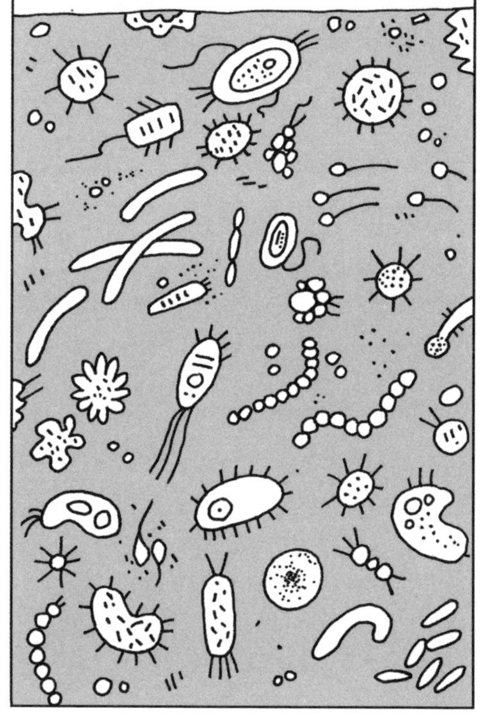

HOYLE AND WICKRAMASINGHE CLAIMED THAT SEVERAL OUTBREAKS OF ILLNESS WERE OF EXTRATERRESTRIAL ORIGIN...

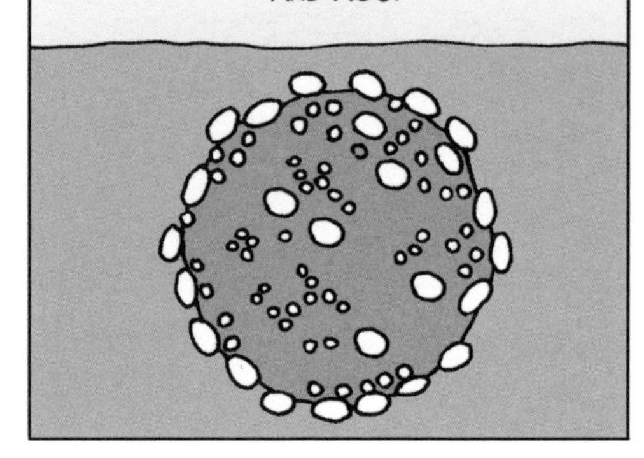

HOYLE WAS UNABLE TO ARRANGE TESTS ON THE ACTUAL SPECIMEN, SO HE UNDERTOOK AN ANALYSIS OF PHOTOGRAPHS OF THE FOSSIL.

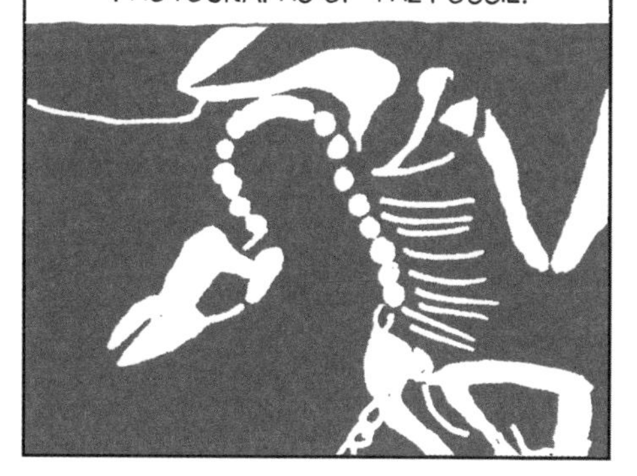

THE NATURAL HISTORY MUSEUM COUNTERED WITH ITS OWN STUDY. IT FOUND THAT MICROCRACKS EXTENDED THROUGHOUT THE FOSSIL AND INTO THE SURROUNDING ROCK. CONTRARY TO WHAT HOYLE CLAIMED, NO MATERIAL HAD BEEN ADDED.

HOYLE'S CONTENTION THAT LIFE HAD DEVELOPED OUT IN SPACE PUT HIM AT ODDS WITH CERTAIN ASPECTS OF THE THEORY OF EVOLUTION.

HE WAS PARTICULARLY SCATHING ABOUT THE ROLE OF CHANCE IN THE EVOLUTIONARY PROCESS.

WHAT IS THE CHANCE THAT AFTER ITS PASSAGE A FULLY ASSEMBLED 747, READY TO FLY, WILL BE FOUND STANDING THERE? SO SMALL AS TO BE NEGLIGIBLE, EVEN IF A TORNADO WERE TO BLOW THROUGH ENOUGH JUNKYARDS TO FILL THE WHOLE UNIVERSE.

THIS IS A COMPLETE MISUNDERSTANDING OF EVOLUTION. CHANCE DOES PLAY A PART, BUT IT'S FAR FROM THE ENTIRE STORY. DNA (DEOXYRIBONUCLEIC ACID) IS THE MOLECULE THAT CARRIES GENETIC INFORMATION FROM GENERATION TO GENERATION.

AN ORGANISM'S DNA AFFECTS EVERY ASPECT OF ITS LIFE - HOW IT LOOKS AND BEHAVES, AND THE FUNCTIONING OF ITS BODY. SO A RANDOM ERROR, A MUTATION, IN THE COPYING OF THIS MOLECULE CAN AFFECT LIVING CREATURES. SOME OF THESE CHANGES ARE BENEFICIAL, WHILE OTHERS ARE NOT.

A MUTATION THAT CAUSED A MOTH'S COLOUR TO DARKEN SO THAT IT BLENDED IN WELL WITH ITS SURROUNDINGS...

WOULD GIVE IT A SURVIVAL ADVANTAGE OVER OTHERS OF ITS KIND.

THE LIGHTER MOTHS, BEING MORE VISIBLE, WOULD BE MORE LIKELY TO BE EATEN BY PREDATORS.

THE DARKER MOTH WOULD PASS ITS GENES AND ASSOCIATED CAMOUFLAGE ON TO ITS DESCENDANTS...

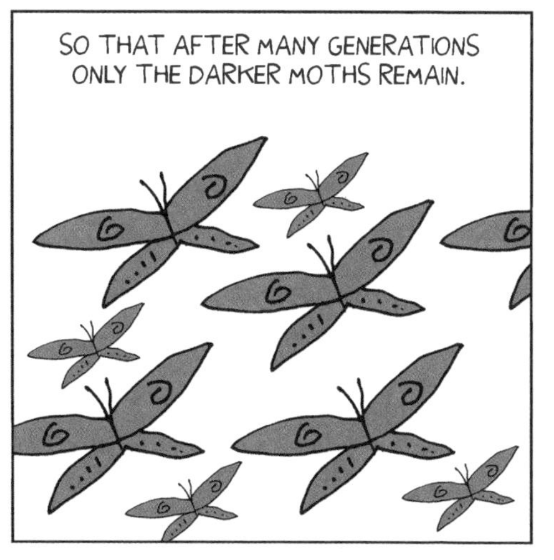
SO THAT AFTER MANY GENERATIONS ONLY THE DARKER MOTHS REMAIN.

THESE SLIGHT ADVANTAGES CAN BUILD UP OVER MILLIONS OF YEARS...

TO CHANGE A CREATURE DRAMATICALLY.

THIS IS NOT JUST CHANCE, BUT THE PROCESS OF NATURAL SELECTION. WHAT WORKS STAYS, AND WHAT DOESN'T WORK FALLS AWAY.

WITHIN THE SCIENTIFIC COMMUNITY THERE ARE DEBATES ABOUT TOPICS SUCH AS THE LEVEL AT WHICH SELECTION OPERATES...

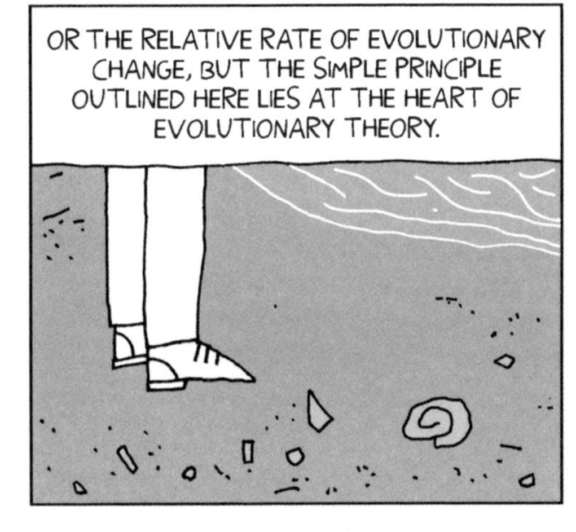
OR THE RELATIVE RATE OF EVOLUTIONARY CHANGE, BUT THE SIMPLE PRINCIPLE OUTLINED HERE LIES AT THE HEART OF EVOLUTIONARY THEORY.

THAT A SCIENTIST OF FRED HOYLE'S CALIBRE WOULD MISUNDERSTAND THIS IS SOMEWHAT SURPRISING.

ALTHOUGH HE WAS AN ATHEIST, CERTAIN EVIDENCE LED HOYLE TO WONDER IF THERE WAS A GUIDING HAND IN THE UNIVERSE.

HE FOUND IT HARD TO BELIEVE THE UNIVERSE HAD RANDOMLY GENERATED THE PRECISE ENERGY LEVEL NECESSARY FOR LIFE-GIVING CARBON.

THE COSMOS LOOKED SUSPICIOUSLY FINE-TUNED TO HIM. TOO MANY OF THE LAWS OF PHYSICS WERE CONVENIENTLY JUST RIGHT.

FOR EXAMPLE, IF THE FORCE OF GRAVITY WERE A FEW PER CENT WEAKER, IT WOULD NOT SQUEEZE AND HEAT THE CENTRE OF THE SUN ENOUGH TO IGNITE THE NUCLEAR REACTION THAT GENERATES SUNLIGHT, BUT IF IT WERE A FEW PER CENT STRONGER, THE TEMPERATURE OF THE SOLAR CORE WOULD HAVE BEEN BOOSTED TO SUCH A DEGREE THAT THE SUN WOULD HAVE BURNED OUT IN LESS THAN A BILLION YEARS. WHICH IS NOT ENOUGH TIME FOR COMPLEX LIFE TO EVOLVE.

JUST ONE IN AN INFINITY OF UNIVERSES THAT ALL HAVE DIFFERENT VALUES FOR THE FUNDAMENTAL LAWS OF PHYSICS.

THE CONCEPT OF MULTIPLE UNIVERSES ENSURING OUR OWN EXISTENCE IS HARD TO GRASP AND IS SOMEWHAT MIND-BLOWING, BUT ESSENTIALLY, IF THE NUMBER OF UNIVERSES IS INFINITE, THEN THEY MUST INCLUDE ALL POSSIBLE VARIATIONS THAT CAN EXIST, MAKING A UNIVERSE WITH THE EXACT REQUIREMENTS FOR OUR KIND OF LIFE INEVITABLE.

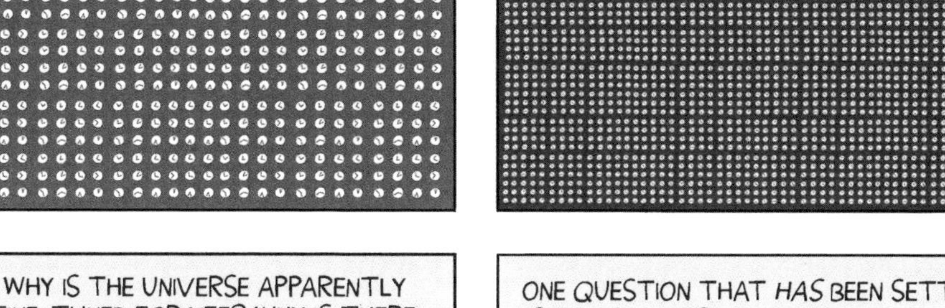
SO IT SHOULD COME AS NO SURPRISE THAT WE FIND OURSELVES LIVING IN ONE.
HOWEVER, THIS IS A HUGELY DISPUTED THEORY IN COSMOLOGY, AND CURRENTLY THERE IS NO WAY OF TESTING WHETHER MULTIPLE UNIVERSES EXIST.

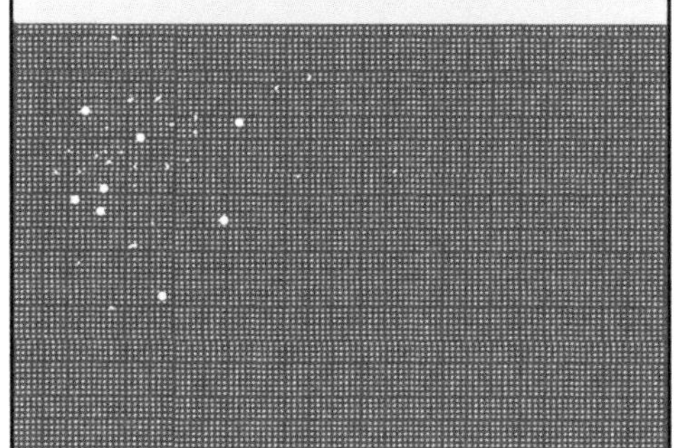
WHY IS THE UNIVERSE APPARENTLY FINE-TUNED FOR LIFE? WHY IS THERE SOMETHING RATHER THAN NOTHING? SUCH DEEP MYSTERIES OF NATURE REMAIN UNEXPLAINED.

ONE QUESTION THAT *HAS* BEEN SETTLED IS THE DECADES-LONG ARGUMENT OVER WHICH OF THE TWO MODELS OF THE UNIVERSE IS CORRECT. WAS IT THE BIG BANG OR THE STEADY STATE?

IN 1964, ARNO PENZIAS AND ROBERT WILSON AT THE BELL TELEPHONE LABORATORIES IN NEW JERSEY...

WERE USING A HORN ANTENNA TO STUDY RADIO WAVES FROM THE MILKY WAY...

WHEN THEY FOUND THEMSELVES REPEATEDLY PICKING UP A NOISE THAT SEEMED TO EMANATE FROM ALL POINTS IN THE SKY.

THINKING IT WAS INTERFERENCE, THEY DID EVERYTHING THEY COULD TO MINIMISE THE NOISE, INCLUDING CLEARING OUT PIGEONS FROM INSIDE THE ANTENNA.

BUT IT WASN'T INTERFERENCE. PENZIAS AND WILSON HAD MADE A SIGNIFICANT DISCOVERY – ONE THAT WOULD WIN THEM THE NOBEL PRIZE.

WHAT THEY HAD DETECTED WAS THE LEFTOVER, COOLED RADIATION FROM THE BIG BANG, WHICH PERVADES THE ENTIRE UNIVERSE.

REGISTERING AT 2.73 DEGREES ABOVE ABSOLUTE ZERO, ITS EXISTENCE HAD BEEN PREDICTED BY GEORGE GAMOW, AMONG OTHERS.

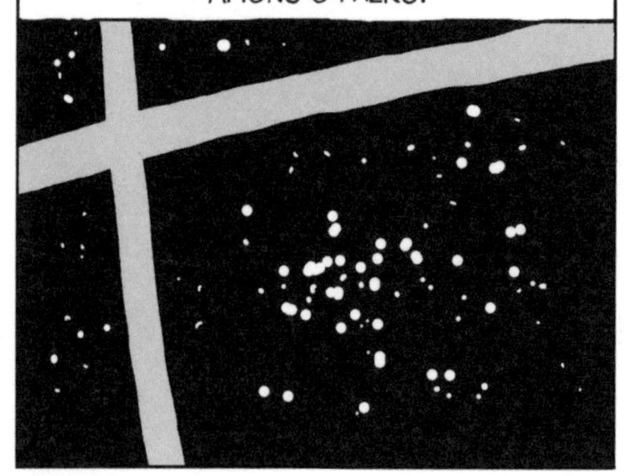

THIS EVIDENCE STRONGLY SUPPORTED THE BIG BANG MODEL OF THE UNIVERSE.

THE COSMIC MICROWAVE BACKGROUND RADIATION WAS MAPPED IN MORE DETAIL BY NASA'S COSMIC BACKGROUND EXPLORER (COBE) IN THE EARLY 1990s, AND BY THE WILKINSON MICROWAVE ANISOTROPY PROBE (WMAP), LAUNCHED IN 2001.

THESE SATELLITES, AND THE EVIDENCE THAT THERE WERE YOUNG GALAXIES AND QUASARS SCATTERED THROUGHOUT THE UNIVERSE, KILLED THE STEADY STATE MODEL. FRED HOYLE, AS OBSTINATE AS EVER, REFUSED TO ACCEPT THIS.

THE REASON WHY SCIENTISTS LIKE BIG BANG IS BECAUSE THEY ARE OVERSHADOWED BY *THE BOOK OF GENESIS*. IT IS DEEP WITHIN THE PSYCHE OF MOST SCIENTISTS TO BELIEVE THE FIRST PAGE OF *GENESIS*.

IT'S POSSIBLE THAT, BY 1985, THE NOBEL PRIZE COMMITTEE CONSIDERED HOYLE UNWORTHY OF THE AWARD NOT JUST BECAUSE OF THE MANY CONTROVERSIES HE WAS INVOLVED IN, BUT ALSO AS A CONSEQUENCE OF HIS NEVER RECANTING HIS BELIEF IN THE STEADY STATE MODEL DESPITE CLEAR EVIDENCE AGAINST IT.

HOYLE MAY HAVE MISSED OUT ON THE TOP PRIZE IN PHYSICS, BUT HE STILL ACHIEVED MUCH IN HIS LIFE. HE WAS APPOINTED THE PLUMIAN PROFESSOR OF ASTRONOMY AT CAMBRIDGE IN 1959...

AND HE BECAME THE FIRST DIRECTOR OF THE CAMBRIDGE INSTITUTE OF OF THEORETICAL ASTRONOMY IN 1967.

HE WAS KNIGHTED IN 1973, BECOMING SIR FRED HOYLE, AT THE AGE OF 57.

ULTIMATELY HE BECAME TIRED OF BATTLING WHAT HE SAW AS THE ABSURD POLITICS OF THE UNIVERSITY. BY 1972 HE'D RESIGNED BOTH HIS POSTS.

HOYLE SPENT THE REST OF HIS LIFE AS A TRAVELLING ASTROPHYSICIST, LECTURING IN UNIVERSITIES AROUND THE WORLD...

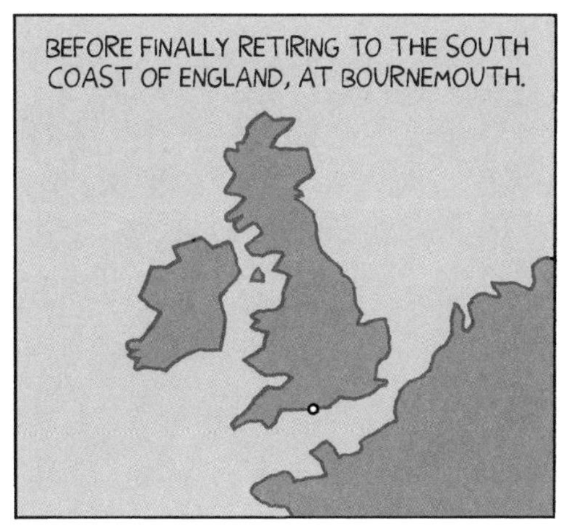
BEFORE FINALLY RETIRING TO THE SOUTH COAST OF ENGLAND, AT BOURNEMOUTH.

IN NOVEMBER 1997, NOW AGED 83, HOYLE DROVE TO HIS CHILDHOOD HOME IN GILSTEAD, WHERE HIS SISTER HAD SPENT HER WHOLE LIFE.

OUT FOR A WALK, HE TOOK A WOODED PATH ON THE UPPER SLOPES OF A STEEP RAVINE, CALLED SHIPLEY GLEN...

WHEN HE SLIPPED...

FALLING TEN FEET AND HITTING HIS HEAD.

WAS THERE SOMETHING MORE TO THIS FALL? HOYLE HAD NO CLEAR MEMORY OF THE ACCIDENT, BUT HIS WALLET WAS EMPTY OF THE 200 POUNDS THAT WAS IN IT WHEN HE'D LEFT THE HOUSE.

FRED HOYLE LIVED INTO THE 21ST CENTURY, CELEBRATING HIS 86TH BIRTHDAY IN JUNE 2001.

SIX WEEKS LATER HE HAD THE LAST OF A SERIES OF STROKES AND DIED ON 20TH AUGUST.

THE AMERICAN ASTRONOMER GEORGE O. ABELL SAID OF HOYLE...
HIS ENTHUSIASM ABOUT ALMOST EVERYTHING IS EXTREMELY CONTAGIOUS.

HE SIMPLY IS THE KIND OF PERSON THAT THINGS OCCUR TO DURING ANY KIND OF CONVERSATION UNDER ALMOST ANY CIRCUMSTANCE.

IT IS FROM SUCH A WEALTH OF IDEAS, SOME OF WHICH ARE WRONG, OTHERS OF WHICH ARE WRONG BUT BRILLIANT...

AND STILL OTHERS OF WHICH ARE BRILLIANT AND RIGHT, THAT SCIENTIFIC PROGRESS IS MADE.
END

## ACKNOWLEDGEMENTS:

As ever I'm indebted to the many writers whose work I drew from while researching *Graphic Science*. So thanks to Joe Jackson, Madison Smartt Bell, Patricia Pierce, Deborah Cadbury, Gary R. Kremer, W. Bernard Carlson, Marc J. Seifer, Robert Lomas, Mott T. Green, Lisa Yount, Simon Mitton, Jane Gregory, and Simon Singh.

Others who helped and supported me in various ways were Nick Abadzis, Gary Barker, Simon Fraser, David Gaffney, Larry Gonick, Katie Green, Luke Foster, Brian Talbot, Mary Talbot, Graham Johnstone, Jonathan Edwards, Louise Evans, Stephen Betts, Ellen Lindner, and my parents.

Thanks also to Candida Lacey and everyone at Myriad Editions. A particularly big thanks must go to my editor Corinne Pearlman for her hard work and patience.

And of course to Bonnie Millard for everything.

**ANTOINE LAVOISIER**

Donovan AL. Antoine-Laurent Lavoisier. *Encyclopedia Britannica.* www.britannica.com/biography/Antoine-Laurent-Lavoisier (accessed 2015 Nov 10).

Marie Paulze Lavoisier Facts. Your Dictionary.com. http://biography.yourdictionary.com/marie-paulze-lavoisier (accessed 2015 Nov 10).

Joseph Priestley and the Discovery of Oxygen. American Chemical Society. 1994 Aug 1. www.acs.org/content/acs/en/education/whatischemistry/landmarks/josephpriestleyoxygen.html (accessed 2015 Nov 23).

French Revolution. *Encyclopedia Britannica.* www.britannica.com/event/French-Revolution (accessed 2015 Nov 24).

Antoine-Laurent Lavoisier. Chemical Heritage Foundation. www.chemheritage.org/historical-profile/antoine-laurent-lavoisier (accessed 2015 Nov 10).

Jackson J. *A World On Fire: A Heretic, An Aristocrat, and the Race to Discover Oxygen.* Penguin: London; 2005.

Smartt Bell M. *Lavoisier in the Year One: The Birth of a New Science in an Age of Revolution* (Great Discoveries series). W.W. Norton & Company: New York and London; 2005, 2nd edn 2006.

**MARY ANNING**

What is a fossil? www.discoveringfossils.co.uk/whatisafossil.htm (accessed 2016 Jan 3).

Mary Anning: Google doodle celebrates the missing woman of geology. *The Guardian.* 2013 Aug 21. www.theguardian.com/science/the-h-word/2013/aug/21/photograph-mary-anning-women-history-geology (accessed 2016 Jan 3).

Pierce P. *Jurassic Mary: Mary Anning and the Primeval Monsters.* The History Press: Stroud; 2014.

Cadbury D. *The Dinosaur Hunters: A True Story of Scientific Rivalry and the Discovery of the Prehistoric World.* Fourth Estate: London; 2000, paperback edn 2010.

**GEORGE WASHINGTON CARVER**

Kremer GR. *George Washington Carver: A Biography.* (Greenwood Biographies series). ABC-CLIO LLC: Santa Barbara; 2011.

George Washington Carver and Other Christians who were Scientists. Georgia Southern University. https://sites.google.com/a/georgiasouthern.edu/etmcmull/george-washington-carver-and-other-christians-who-were-scientists (accessed 2016 Dec 7).

Cannon W. A True Renaissance Man. *American Scientist.* www.americanscientist.org/bookshelf/pub/a-true-renaissance-man (accessed 2016 Dec 7).

Harris P. How the end of slavery led to starvation and death for millions of black Americans. *The Guardian.* 2012 June 16. www.theguardian.com/world/2012/jun/16/slavery-starvation-civil-war (accessed 2016 Dec 18).

Booker T. Washington. Biography.com. www.biography.com/people/booker-t-washington-9524663 (accessed 2016 Dec 16).

Booker T. and W.E.B.: The debate between W.E.B Du Bois and Booker T. Washington. PBS *Frontline.* www.pbs.org/wgbh/pages/frontline/shows/race/etc/road.html (accessed 2016 Dec 16).

Who Invented Peanut Butter? National Peanut Board. http://nationalpeanutboard.org/peanut-info/who-invented-peanut-butter.htm (accessed 2016 Dec 25).

The George Washington Carver Museum. Wikipedia. https://en.wikipedia.org/wiki/The_George_Washington_Carver_Museum (accessed 2016 Dec 27).

George Washington Carver. *Encyclopedia Britannica.* www.britannica.com/biography/George-Washington-Carver (accessed 2016 Dec 7).

**NIKOLA TESLA**

Carlson WB. *Tesla: Inventor of the Electrical Age.* Princeton University Press: Princeton and Oxford; 2013.

Seifer MJ. *Wizard: The Life and Times of Nikola Tesla: Biography of a Genius.* Citadel Press: New York; 1996.

Lomas R. *The Man Who Invented the Twentieth Century: Nikola Tesla, Forgotten Genius of Electricity.* Headline: London; 1999, reissued via CreateSpace 2013.

Tesla N. *My Inventions and Other Writings* (Penguin Classics series). Penguin: London; 2012.

AC Power History and Timeline: The History of Alternating Current. Edison Tech Centre. http://edisontechcenter.org/AC-PowerHistory.html (accessed 2016 July 12).

Galileo Ferraris. Encyclopedia.com. www.encyclopedia.com/people/science-and-technology/physics-biographies/galileo-ferraris (accessed 2016 July 23).

Edison's 1882 Pearl Street Station. Alevo. http://alevo.com/edisons-1882-pearl-street-station/ (accessed 2016 July 27).

King G. Edison vs. Westinghouse: A Shocking Rivalry. *Smithsonian.* 2011 Oct 11. www.smithsonianmag.com/history/edison-vs-westinghouse-a-shocking-rivalry-102146036 (accessed 2016 July 29).

War of currents. Inventions and Research of Nikola Tesla. https://teslaresearch.jimdo.com/war-of-currents (accessed 2016 July 27).

Rouse M. Frequency. *WhatIs.com*. 2010 Mar. http://whatis.techtarget.com/definition/frequency (accessed 2016 July 27).

"Death ray". Inventions and Research of Nikola Tesla. https://teslaresearch.jimdo.com/death-ray (accessed 2016 Aug 3).

## ALFRED WEGENER

Greene MT. *Alfred Wegener: Science, Exploration, and the Theory of Continental Drift*. Johns Hopkins University Press: Baltimore; 2015.

Yount L. *Alfred Wegener: Creator of the Continental Drift Theory* (Makers of Modern Science series). Chelsea House Publishers: New York; 2009.

Conniff R. When Continental Drift Was Considered Pseudoscience. *Smithsonian*. 2012 June. www.smithsonianmag.com/science-nature/when-continental-drift-was-considered-pseudoscience-90353214 (accessed 2016 Sept 3).

The Danmark Expedition 1906–1908. Environment & Society Portal. www.environmentandsociety.org/exhibitions/wegener-diaries/expedition1 (accessed 2016 Sept 3).

Plate Tectonics. BBC website. www.bbc.co.uk/science/earth/surface_and_interior/plate_tectonics (accessed 2016 Sept 8).

Oskin B. What Is Plate Tectonics? Live Science. 2016 Mar 21. www.livescience.com/37706-what-is-plate-tectonics.html (accessed 2016 Sept 21).

Wegener, Alfred Lothar. Encyclopedia.com. www.encyclopedia.com/people/science-and-technology/geology-and-oceanography-biographies/alfred-lothar-wegener (accessed 2016 Sept 3).

## JOCELYN BELL BURNELL

Gold L. Discoverer of pulsars (aka Little Green Men) reflects on the process of discovery and being a female pioneer. *Cornell Chronicle*. 2006 July 6. www.news.cornell.edu/stories/2006/07/jocelyn-bell-burnell-reflects-discovery-pulsars (accessed 2015 Oct 10).

Dame Jocelyn Bell Burnell. *The Life Scientific*. BBC Radio 4. 2011 Oct 25. http://www.bbc.co.uk/programmes/b016812j (accessed 2015 Oct 10).

Edmonds J. Tutchell E. *Man-Made: Why So Few Women Are in Positions of Power*. Gower Publishing: Farnham; 2015; 2nd edn Routledge: New York; 2016.

Interplanetary Scintillation Array. Wikipedia. https://en.wikipedia.org/wiki/Interplanetary_Scintillation_Array (accessed 2015 Oct 16).

Cain F. What Is a Pulsar? *Universe Today*. Updated 2017 Mar 16. https://www.universetoday.com/25376/pulsars/ (accessed 2015 Oct 15).

Professor Dame Jocelyn Bell Burnell. *Woman's Hour* Profiles. http://www.bbc.co.uk/programmes/profiles/1fF5G4ngxbm5KsWQTGDYX1/professor-dame-jocelyn-bell-burnell (accessed 2015 Oct 10).

Anthony Michaelis obituary. *The Telegraph*. 2008 Mar 28. www.telegraph.co.uk/news/obituaries/1583056/Anthony-Michaelis.html (accessed 2017 Feb 10).

## FRED HOYLE

Singh S. *Big Bang: The Most Important Scientific Discovery of All Time and Why You Need to Know About It*. Harper Perennial: Hammersmith; 2005.

Mitton S. *Fred Hoyle: A Life in Science*. Cambridge University Press: Cambridge; 2011.

Gregory J. *Fred Hoyle's Universe*. Oxford University Press: Oxford; 2005.

Lovell B. Sir Fred Hoyle obituary. *The Guardian*. 2001 Aug 23. www.theguardian.com/news/2001/aug/23/guardianobituaries.spaceexploration (accessed 2016 Oct 1).

*A for Andromeda*. Wikipedia. https://en.wikipedia.org/wiki/A_for_Andromeda (accessed 2016 Oct 21).

Lezard N. *The Black Cloud* by Fred Hoyle – review. *The Guardian*. 2010 Oct 23. www.theguardian.com/books/2010/oct/23/black-cloud-fred-hoyle-review (accessed 2016 Oct 21).

Clarke Dr NM. Life, Bent Chains, and the Anthropic Principle. University of Birmingham Nuclear Physics Group. 1999. www.np.ph.bham.ac.uk/history/nucleosynthesis (accessed 2016 Nov 2).

Siegel E. Starts with a Bang! The Last 100 Years: 1950s & The Tragedy of Fred Hoyle. ScienceBlogs. 2009 June 23. http://scienceblogs.com/startswithabang/2009/06/23/the-last-100-years-1950s-the-t (accessed 2016 Oct 20).

William A. Fowler – Facts. Nobel prize.org. www.nobelprize.org/nobel_prizes/physics/laureates/1983/fowler-facts.html (accessed 2016 Nov 4).

Aron J. Stars burning strangely make life in the multiverse more likely. *New Scientist*. 2016 Sept 1. www.newscientist.com/article/2104223-stars-burning-strangely-make-life-in-the-multiverse-more-likely (accessed 2016 Nov 9).

Chown M. Why the universe wasn't fine-tuned for life. *New Scientist*. 2011 June 14. www.newscientist.com/blogs/culturelab/2011/06/why-the-universe-wasnt-fine-tuned-for-life.html (accessed 2016 Nov 8).

Cosmic Microwave Background Radiation. BBC website. www.bbc.co.uk/science/space/universe/sights/cosmic_microwave_background_radiation (accessed 2016 Nov 19).

## MORE BY DARRYL CUNNINGHAM FROM MYRIAD

A graphic milestone of investigative reporting, Cunningham's essays explode the lies, hoaxes and scams of popular science, debunking media myths and decoding some of today's most fiercely-debated issues: climate change, electroconvulsive therapy, fracking, the moon landing, the MMR (Measles, Mumps and Rubella) vaccine, homeopathy, chiropractic, evolution, and science denialism.

ISBN 978-1-908434-36-4
eISBN 978-1-908434-62-3

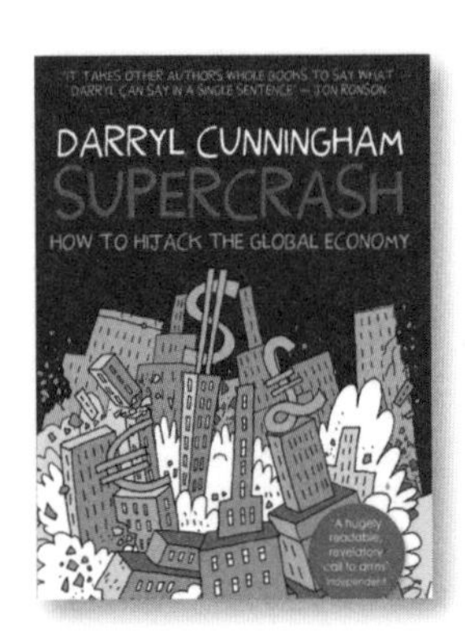


Darryl Cunningham takes us to the heart of the economic crisis, showing how the Neo-Cons hijacked the debate and led the way to a world dominated by the market. He traces the roots of the financial crisis of 2008 to the domination of right-wing policies and the people who created them, drawing a fascinating portrait of the New Right and the charismatic Ayn Rand. He examines the neurological basis of political thinking, and asks why it is so difficult for us to change our minds – even when faced with powerful evidence that a certain course of action is not working.

ISBN 978-1-908434-43-2
eISBN 978-1-908434-73-9